Process Oriented Guided Inquiry Learning (POGIL)

ACS SYMPOSIUM SERIES **994**

Process Oriented Guided Inquiry Learning (POGIL)

Richard S. Moog, Editor
Franklin and Marshall College

James N. Spencer, Editor
Franklin and Marshall College

Sponsored by the
ACS Division of Chemical Education, Inc.

American Chemical Society, Washington, DC

Library of Congress Cataloging-in-Publication Data

Process-oriented guided inquiry learning (POGIL) / Richard S. Moog, editor, James N. Spencer, editor ; sponsored by the Division of Chemical Education, Inc.

p. cm.—(ACS symposium series ; 994)

Includes bibliographical references and index.

ISBN 978–0–8412–7423–5 (alk. paper)

1. Process-oriented inquiry learning. 2. Chemistry—Study and teaching (Higher).
3. Chemistry—Study and teaching (Secondary).

I. Moog, Richard Samuel. II. Spencer, James N. (James Nelson), 1941- III. American Chemical Society. Division of Chemical Education, Inc.

QD49.6.P76P76 2008
540.71′2—dc22

2008060009

The paper used in this publication meets the minimum requirements of American National Standard for Information Sciences—Permanence of Paper for Printed Library Materials, ANSI Z39.48–1984.

Distributed by Oxford University Press

PRINTED IN THE UNITED STATES OF AMERICA

Foreword

The ACS Symposium Series was first published in 1974 to provide a mechanism for publishing symposia quickly in book form. The purpose of the series is to publish timely, comprehensive books developed from ACS sponsored symposia based on current scientific research. Occasionally, books are developed from symposia sponsored by other organizations when the topic is of keen interest to the chemistry audience.

Before agreeing to publish a book, the proposed table of contents is reviewed for appropriate and comprehensive coverage and for interest to the audience. Some papers may be excluded to better focus the book; others may be added to provide comprehensiveness. When appropriate, overview or introductory chapters are added. Drafts of chapters are peer-reviewed prior to final acceptance or rejection, and manuscripts are prepared in camera-ready format.

As a rule, only original research papers and original review papers are included in the volumes. Verbatim reproductions of previously published papers are not accepted.

ACS Books Department

Contents

Indexes

Acknowledgments

This volume is the outgrowth of two symposia on Process-Oriented Guided Inquiry Learning (POGIL) held at consecutive American Chemical Society (ACS) meetings: August 30–September 1, 2005 in Washington, D.C. and March 26–28, 2006 in Atlanta, Georgia. The editors thank Dara Moore for first inviting the POGIL Project to submit a proposal for a contribution to the ACS Symposium Series Books. Perhaps most importantly, the editors thank Dara for her continuous encouragement, helpful suggestions, and patience.

Without the authors and those who reviewed the book chapters, no book would have been possible. The editors acknowledge and thank the following individuals for providing chapter reviews on this project:

David Parkin	Adelphi University, Garden City, NY
Anne Falke	Worcester State College, Worcester, MA
R. Daniel Libby	Moravian College, Bethlehem, PA
Robin Lasey	Arkansas Tech, Russelville, AR
Ashley Mahoney	Bethel University, St. Paul, MN
Kelly Butler	Chestnut Hill College, Philadelphia, PA
Deborah Pomeroy	Arcadia University, Glenside, PA
Andrea Martin	Widener University, Chester, PA
Alex Grushow	Rider University, Lawrenceville, NJ
Maureen Scharberg	San Jose State University, San Jose, CA
Candice Foley	Suffolk Community College, Selden, NY
Susan Shadle	Boise State University, Boise, ID
Gail Webster	Guilford College, Greensboro, NC

The National Science Foundation provided support to the POGIL Project through the following grants:

DUE 0231120
Principal Investigators:

Richard Moog	Franklin and Marshall College, Lancaster, PA
Frank Creegan	Washington College, Chestertown, MD
David Hanson	Stony Brook University, Stony Brook, NY
James N. Spencer	Franklin and Marshall College, Lancaster, PA
Andrei Straumanis	College of Charleston, Charleston, SC

DUE 0618746, 0618758, 0618800
Principal Investigators:

Richard Moog	Franklin and Marshall College, Lancaster, PA
Diane Bunce	The Catholic University, Washington , DC
Jennifer Lewis	University of South Florida, Tampa, FL

The MID Project generously invited POGIL to participate in the dissemination of new teaching strategies developed by various instructors across the country.

POGIL headquarters are currently located at Franklin and Marshall College, and we wish to acknowledge the support of the institution in many ways. The project now occupies two dedicated and renovated rooms in the middle of campus.

We gratefully acknowledge the work of the POGIL staff, Stephanie, Stephen, Ellen and Cindy, without whom the book could not have been completed and the POGIL Project could not function.

Process Oriented Guided Inquiry Learning (POGIL)

Chapter 1

POGIL: An Overview

Richard S. Moog and James N. Spencer

Department of Chemistry, Franklin and Marshall College, P.O. Box 3003, Lancaster, PA 17604–3003

POGIL (Process-Oriented Guided Inquiry Learning) is a student-centered, research-based pedagogic strategy that has been used effectively in chemistry classrooms at all levels in colleges and high schools throughout the country. This approach is built on the foundational work of many others in the areas of cognitive development, cooperative learning, and instructional design. In addition, the reform efforts in science curriculum and pedagogy of the late twentieth century, particularly those in chemistry, were instrumental in laying the groundwork for POGIL and the POGIL Project, a national faculty development effort.

Process-Oriented Guided Inquiry Learning (POGIL) is a research-based instructional philosophy and methodology based on our current understanding of how students learn best. In a POGIL learning environment, students are actively engaged in mastering the concepts and content of a discipline; at the same time, they are developing important learning skills by working in self-managed teams on guided inquiry activities designed specifically for this purpose and environment. In this chapter, a brief introduction to POGIL is provided, along with its philosophical basis and historical context. Throughout this narrative, reference will be made to the various chapters of this volume containing more detailed information about each topic. This chapter ends with a brief description of the POGIL Project, a national professional development effort whose focus is promoting a student-centered, research-based approach to undergraduate science instruction through the implementation of POGIL.

Issues Concerning Traditional Instruction

In a traditional teacher-centered classroom, the vast majority of class time is spent with the teacher talking. There is often relatively little student-teacher interaction, and similarly almost no student-student give and take. The instructor lectures, explains concepts, presents analogies, provides answers, and may demonstrate step-by-step procedures for how to solve various exercises. Although this can be an efficient way to present information, it does not necessarily provide a learning environment in which students learn effectively and develop crucial thinking skills (*1*). Eric Mazur (*2*) has suggested that one possible explanation for the survival of these techniques is that even experienced teachers may be misled as to whether students are truly learning concepts rather than memorizing algorithms. A more effective learning environment is one in which the students can actively engage, an environment in which there is something for students to do (*3*). A common finding in research on how people learn is that telling is not teaching; an idea cannot be transferred intact from the head of the instructor to the head of the student (*4*). In order to help students develop appropriate understanding, it is necessary to know what is going on in the student's mind. Thus, instructors need to put themselves in a position to be so informed. This new paradigm is put succinctly by Elmore (*5*): "Knowledge results only through active participation in its construction. Students teach each other and they teach the instructor by revealing their understanding of the subject."

This perspective suggests that the instructional focus should be on the *activity* of the *students* rather than the *presentation* of the *instructor*. This is the essence of a student-centered classroom. The role of the instructor is one of a facilitator of learning, asking probing questions to help guide the students to develop understanding, and addressing misconceptions or misunderstanding. Many of these ideas are discussed in more depth and expanded upon in the following two chapters where Hanson (Chapter 2) and Lamba (Chapter 3) describe related cognitive models that serve as a basis for POGIL. In Chapter 4, Bressette describes his dissatisfaction with student conceptual understanding in his traditional classroom, and outlines the process by which he transformed that classroom into a student-centered one. Bunce, in Chapter 9, further addresses issues concerning why instructors consider change and the barriers and support needed by them.

What Is POGIL?

The fundamental principles of POGIL have been described in detail in previous publications (*6-8*). The description that follows is based closely on the presentation given in a recent article in *Metropolitan Universities Journal* (*8*).

Because POGIL is a student-centered instructional approach, in a typical POGIL classroom or laboratory, students work in small groups with the instructor acting as a facilitator. The student groups use specially designed activities that generally follow a learning cycle paradigm (*see below*). These activities are designed to have three key characteristics; they

- are designed for use with self-managed teams that employ the instructor as a facilitator of learning rather than as a source of information;
- guide students through an exploration to construct understanding;
- use discipline content to facilitate the development of important process skills including higher-level thinking and the ability to learn and to apply knowledge in new contexts.

The POGIL approach has two broad aims: to develop content mastery through student construction of their own understanding, and to develop and improve important learning skills such as information processing, oral and written communication, critical thinking, problem solving, and metacognition and assessment. The essential components of POGIL – active engagement of students through group learning, guided inquiry materials based on the learning cycle paradigm, and a focus on process skill development – are each described in more detail below.

Active Engagement and Group Learning

Active student engagement and social interaction are now recognized as being essential for most students to gain true understanding and long-term retention. A large amount of evidence (*1*) indicates that individuals must take an active role in what they are trying to learn (including asking questions, engaging in dialogue and discussion, recreating ideas in their own minds, and manipulating and transforming them as needed in new contexts) in order for the learning to be truly lasting. In addition, recent results from research on learning shows that knowledge acquisition and its application are essentially social acts (*9*). These ideas are consistent with the substantial literature on cooperative learning and its effectiveness. When students work together to construct understanding, debate and discuss different ideas to resolve them, and share ideas and strategies, their performance as individuals (for example, on examinations) improves. An excellent introduction to the use of small groups and their effectiveness in science classrooms has been presented recently by Cooper (*10*). In Chapter 8 of this volume, Daubenmire and Bunce report on the dynamics of groups within a typical POGIL classroom.

The Learning Cycle

The learning cycle is an inquiry strategy for teaching and learning that is based on constructivist principles. Abraham (*11*) has recently presented an exellent overview of this approach and its effectiveness; we briefly summarize the main points here. Lawson (*12*) describes a learning cycle consisting of three phases:

1. An "Exploration" phase in which a pattern of regularity in the environment or data (collected by the students, or presented to them directly) is sought. Students generate hypotheses and test them in an attempt to explain or understand this information.
2. A "Concept Invention" or "Term Introduction" phase in which a concept is developed from patterns in the data and a new term is introduced to refer to these newly-identified trends or patterns. By having the "Term Introduction" phase follow the "Exploration" phase, new terms are introduced at a point when the student has already constructed her own understanding of the concept to which the term is then attached. This is in contrast to a typical textbook (or lecture) presentation in which terms are frequently presented or defined before examples of their use are given.
3. An "Application" phase in which the just-developed concept is applied in new situations. This phase is intended to generalize the concept's meaning and applicability, frequently requiring deductive reasoning skills.

With this structure, a learning cycle experience guides students to develop concepts for themselves, promoting a sense of ownership and participation, and providing epistemological insight into the nature of scientific inquiry.

A Typical POGIL Activity

To help clarify how the POGIL approach differs from a more traditional lecture or textbook presentation, a portion of a typical POGIL activity for a General Chemistry class (*13*) is described in some detail below. The content includes the basic ideas concerning the components of an atom. Typically, a lecturer would tell the students that the constituents of atoms are protons, neutrons, and electrons, and that the number of protons in the atom is known as the "atomic number" and determines the atom's identity. In the POGIL activity, the approach to this content is markedly different. The activity begins with a series of diagrams providing examples of a number of atoms and ions, with the corresponding element identified along with the number and location of the protons, neutrons, and electrons in each. Through a series of guiding questions, the students are led to recognize that all of the atoms with the same number of protons are identified as the same element. They also are asked to determine the significance of the number 6 with the number on the periodic table that identifies carbon, and are able to conclude that they are the same. Only at this point, as this concept is developed, is the term "atomic number" used to describe the number of protons in one atom of a given element. Thus, an "exploration" of the information presented in the diagrams allows each student to develop the concept that the number of protons determines the identity of an element; the term "atomic number" is introduced following this construction. The "application" of this concept entails using the periodic table to identify the number of protons in other elements.

Key Aspects of POGIL Activity Design

There are two crucial aspects to the design of the POGIL activity. First, sufficient appropriate information must be provided for the initial "Exploration" so that students are able to develop the desired concepts. Second, the guiding questions must be sequenced in a carefully constructed manner so that not only do students reach the appropriate conclusion, but at the same time various process and learning skills are implemented and developed. Typically, the first few questions build on students' prior knowledge and direct attention to the information provided in the model. This is followed by questions designed to help promote the recognition of relationships and patterns in the data, leading toward some concept development. The final questions may involve applying the concepts to new situations and generalizing students' new knowledge and understanding.

Several chapters throughout this volume provide additional details concerning the design of POGIL activities. In Chapter 13, Moog and Spencer provide a detailed description of the structure of a POGIL activity for physical chemistry, and Garoutte discusses an activity for an allied health course in Chapter 11. The use of this approach in a non-science majors course is presented by Lees in Chapter 15, and Creegan and Lamba describe the application of POGIL principles to laboratory experiences in Chapter 16.

A Focus on Process

One of the most important aspects of implementing POGIL is the recognition that there are significant student learning outcomes that are independent of the specific course content. The stated mission of undergraduate education at the vast majority of institutions in the United States is to produce independent life-long learners who will lead meaningful lives and be contributors to society. Yet, for most faculty members, particularly those in the sciences, consciously and intentionally working toward this goal is not generally part of the everyday instructional planning process. In contrast, a POGIL approach specifically places an emphasis on the development of process skills that will help achieve these goals – or whatever other process-oriented learning outcomes the instructor has chosen for that day and/or course, Thus, within the POGIL philosophy, the development of process skills (e.g. information processing, critical thinking, communication, assessment) is a specific and intentional focus of the classroom implementation; improving these skills will not only complement and enhance the mastery of course content for the student, but will also help achieve the overall goals of the institution. The inclusion of this focus on the development of process skills is due in large part to the work of Dan Apple of Pacific Crest (*14, 15*).

For many instructors, intentionally focussing on process skill development can be difficult, in no small part because it is often a very new idea. Finding a way to appropriately measure the effectiveness of an attempt to develop process skills can seem even more daunting. In Chapter 7, Minderhout and Loertscher describe how the development of a facilitation plan for each classroom experience can help the instructor focus on the particular process skills that the students should be working on in class that day. Bauer and Cole provide suggestions for possible ways to assess various aspects of a POGIL implementation in Chapter 18. In addition, a study of students' perceptions of their own development of process skills is described by Straumanis and Simons in Chapter 19. In Chapter 20, Perry and Wright report on the use of an ACS exam to compare traditional and POGIL instruction.

How Is POGIL Implemented?

POGIL can be implemented in a variety of ways, depending on numerous factors such as class size, the nature of the teaching space, and instructor preferences. Successful models include replacing essentially all lectures with POGIL sessions (*13*), replacing one lecture session each week with a POGIL session (*16*) , and implementing POGIL in standard recitation sessions at a large university (*17*). Still, the common features of any POGIL implementation include:

- students working in small groups (generally 3 or 4), typically with assigned roles;
- activities that have been specifically and carefully crafted – usually based on the learning cycle paradigm - to be used in this context;

- an instructor who is not predominantly lecturing, but rather serving as a facilitator of student learning.

Several chapters in this volume describe examples of how POGIL may be implemented in various courses and settings. Padwa, Hanson and Trout describe issues related to the use of POGIL in high school classrooms in Chapter 10. In Chapter 6, Yezierski and coauthors provide a general discussion of POGIL implementation in large classes, and in Chapter 12 Hunnicutt and Ruder present details concerning their implementations in large general chemistry and organic chemistry classes at Virginia Commonwealth University. Chapters 11 and 13 provide examples of POGIL as the major instructional paradigm for the allied health (GOB) and the physical chemistry course. Two different uses of tablet PCs in a POGIL classroom are presented by Mewhinney and Zuckerman in Chapter 14, followed by Lees' description of POGIL as an important component of a non-science majors' course in Chapter 15. Finally, Creegan and Lamba (Chapter 16) and Van Bramer and Martin (Chapter 17) discuss issues related to implementing a POGIL approach in the laboratory component of chemistry courses.

Theoretical and Methodological Foundations of POGIL

POGIL as an instructional paradigm is based on the ideas, research, and creativity of a large number of people working over the past 100 years. Describing the contributions of all who have provided important insights and ideas upon which the POGIL instructional model has been built is truly an impossible task. Here we provide a very brief account of a few of the most significant contributors.

In the early twentieth century, Dewey established the groundwork for student involvement in the classroom (*18*). In the years that followed, Vygotsky (*19*), Ausubel (*20*) and Piaget (*21*) contributed substantially to our understanding of the learning process and laid the foundation for a constructivist model for learning. In Chapter 5, Libby discusses how Piaget's ideas influenced his development as an instructor, and how that development eventually led him to his current POGIL implementation. In the late 1950s and early 1960s, Karplus applied these ideas to the development of the learning cycle (described in some detail previously) as a paradigm for elementary school science instruction (*22*). Later, Lawson and his coworkers extended this approach to higher levels of instruction (*12*), and Abraham has written extensively on this topic since that time (*11*). Abraham and Pavelich's laboratory manual, *Inquiries in Chemistry* (*23*), first published in 1972, was an important effort in bringing the ideas of inquiry in the teaching laboratory to the attention of college-level instructors.

The development of effective practices for the implementation of cooperative learning in the classroom by Johnson and Johnson (*3*) and Slavin

(*24, 25*) provided a strong foundation for the use of student learning groups in a variety of contexts. In the 1970s, Treisman's work with minority students in calculus classes at the University of California - Berkeley, provided empirical evidence for both the effectiveness of group work in developing content mastery, and also the importance of social interactions as part of an effective academic experience (*26*). Since the mid-1990s, Felder, a chemical engineer at North Carolina State University, has been a strong advocate of cooperative learning approaches in engineering courses, and has undertaken longitudinal studies of student performance and retention to demonstrate the effectiveness of the approach (*27-29*). More recently, and specifically in the context of chemistry education, Johnstone (*4*) has discussed cognitive models of information processing and knowledge acquisition that provide substantial insight into how students learn. In Chapters 2 and 3, Johnstone's model is discussed in more detail. Hanson (Chapter 2) expands on the model to more fully account for the learner's processing of information, and Lamba (Chapter 3) describes the implications for instruction that are implied by the model.

Curricular Reform Efforts Lay the Groundwork for POGIL

Many of the science curriculum reform efforts of the late twentieth century – particularly those in chemistry - were crucial in providing a context for the development of POGIL, and for its acceptance as a viable and acceptable method of science instruction. In the late 1950s, the use of inquiry as an instructional strategy for high schools was supported by the National Science Foundation. Two chemistry curricula were developed: the Chemical Bond Approach (CBA) and Chemical Educational Materials Study (CHEM Study). Although the CBA approach is closer to what is generally accepted today as inquiry, particularly in its open-ended laboratory program, it never gained a large following (*30*). CHEMStudy did influence secondary instruction, but was not based on inquiry principles. The major thrust was to introduce modern concepts into the chemistry curriculum (*31*).

In the 1980s, the chemistry faculty at the College of the Holy Cross followed the pioneering work of Abraham and Pavelich (*11*) with the development of an inquiry approach to teaching chemistry that put laboratory work at the center of the educational experience for chemistry courses. The Holy Cross faculty commented that traditional instructional approaches focused on the results of the scientific method of inquiry rather than on the process itself. Noting that it is actually the process of discovery that characterizes science, they developed an approach in which students collected and analyzed data in the laboratory that would enable them to "discover" an important principle or concept through the pooling of data, which was then discussed and built upon during the lecture sessions that followed the experimentation later in the week. This approach was first referred to as "Discovery Chemistry" and later termed

"Guided Inquiry" to better reflect the role of the instructor and the structure of the laboratory experiences in guiding the students in their investigations (*32*). An outgrowth of this effort was the establishment in 1993 of the Middle Atlantic Discovery Chemistry Project (MADCP), a consortium of institutions in the region which built on the ideas of the Holy Cross faculty to develop guided inquiry experiments for general chemistry and organic chemistry courses (*33*). The criteria that this group developed for these experiments formed the basis for the guidelines used for POGIL laboratory projects.

In the early 1990s, the Division of Chemical Education of the American Chemical Society formed a Task Force on the General Chemistry Curriculum with a charge to make recommendations for change in the general chemistry course. The Task Force, with the publication *New Directions for General Chemistry* (*34*), provided information about curricular projects and instructional strategies to implement curricular change. Among the recommendations were to reform the general chemistry curriculum, particularly with respect to the number of topics taught, to make use of what had been learned about how people learn, and to alter the process goals of the course. The broad national discussion of the issues surrounding changing the way that general chemistry is taught was important in stimulating faculty members to consider the possibility of substantial alterations in their classrooms practices in order to improve student learning. The publication of Herron's *The Chemistry Classroom* in 1996 (*35*) not only provided substantive insights and suggestions on chemistry instruction but also raised the awareness of these issues even more broadly.

A number of other important developments with respect to reform of science education during this time period have also been influential in providing a context for the development of POGIL. Project Kaleidoscope (PKAL) (*36*), an organization dedicated to bringing about systemic change and institutionalizing reform in undergraduate STEM teaching and learning, was founded in 1989. PKAL's efforts raised awareness of important teaching and learning issues in undergraduate STEM education, including those related to the design of science buildings and the learning spaces within them. One example of the type of reform in undergraduate science education that PKAL promotes is Problem-Based Learning (PBL), which originated in medical school education as a replacement for large lecture classes (*37*). Beginning in the 1980s, PBL was adapted for college classes in a variety of disciplines (*38-40*), with groups of students working cooperatively on open-ended problems that are generally based on real-world situations.

Significant efforts in physics education in the past twenty years include the implementation of Workshop Physics (*41*) in the late 1980s, the development of Just In Time Teaching (*42*) in the 1990s, and more recently the SCALE UP project (*43*) at North Carolina State University. All of these efforts brought are focused on promoting a more student-centered instructional paradigm. In addition, the development of the Force Concept Inventory (*44*) and its widespread use throughout the physics education community has focused attention on student development of conceptual understanding of scientific concepts rather than merely the ability to provide correct numerical answers to examination questions.

Within the undergraduate chemistry education community, groundbreaking efforts were also taking place, in addition to the work of the ACS Task Force on General Chemisty. In 1994 and 1995, the National Science Foundation awarded several large Systemic Change Initiative grants to stimulate reform in undergraduate chemistry education:

- Establishing New Traditions (DUE-9455928), centered at the University of Wisconsin (*45*)
- ChemLinks (DUE-9455918), centered at Beloit College; and Modular Chemistry Consortium (DUE-9455924), centered at University of California – Berkeley, which joined forces to become Chem Connections (*46*)
- Workshop Chemistry (which has become Peer Led Team Learning or PLTL), centered at City College of New York (*47*)
- Molecular Science (DUE-9555605), centered at University of California – Los Angeles (*48*)

All of these projects advocated a movement toward more student involvement in the learning process, albeit in a variety of ways. Dissemination of these ideas continued through 2004 through the Multi-Initiative Dissemination (MID) Project, continuing to raise the awareness of faculty throughout the country to the benefits of more student-centered approaches to chemistry instruction (*49*). Several of the Principle Investigators on the first POGIL NSF grant had been members of the New Traditions project, and presented many of the fundamental ideas of POGIL through both the New Traditions and MID workshops. In addition, many of the ideas from the other projects have been integrated with POGIL in very effective ways. Many instructors combine POGIL with Calibrated Peer Review, a web-based method for peer-review of writing assignments that was one of the important outcomes of the Molecular Science Initiative. A combination of PLTL with POGIL, referred to as Peer Led Guided Inquiry, has been shown to be a powerful combination (*16*). The use of real-world case studies as a basis for developing and understanding chemistry concepts as developed by Chem Connections was a prelude to the current efforts to develop POGIL materials that include a real-world context, known as POGIL - IC (POGIL in Context, NSF DUE-0632957, 0633231, 0633191). Thus, all of these projects played a pivotal role in both creating a national context for further student-centered reform in chemistry education and providing effective models and approaches to student learning that are complementary and compatible with implementing a POGIL approach in the classroom and laboratory.

The POGIL Project

The POGIL Project provides a mechanism for the further development and dissemination of the POGIL approach to teaching, with substantial financial support provided by the National Science Foundation (DUE 0231120, 0618746, 0618758, 0618800). The focus of the project is on faculty development, helping instructors at both the undergraduate and high school levels find ways to achieve the goals that they have for their students through more student-centered approaches to instruction. The Project runs numerous workshops annually across the country. Additional information, including up-to-date lists of upcoming workshops, is available from the POGIL website, http://www.pogil.org.

References

1. *How People Learn;* Bransford, J. D.; Brown, A. L.; Cocking, R. R., Eds.; National Academy Press: Washington, DC, 1999.
2. Mazur, E. *Peer Instruction;* Prentice Hall: Upper Saddle River, NJ, 1997.
3. Johnson, D. W.; Johnson, R. T.; Smith, K. A. *Active Learning: Cooperation in the College Classroom;* Interaction Book Company: Edina, MN, 1991.
4. Johnstone, A. H. *J. Chem. Educ.* **1997**, *74*, 262-268.
5. Elmore, R. F. *Foreward,* In *Education for Judgement;* Christensen, C.R., Garvin, D.A., and Sweet, A., Eds.; Harvard Business School: Boston, MA, 1991.
6. Hanson, D. *Instructor's Guide to Process-Oriented Guided Inquiry Learning*; Pacific Crest: Lisle, IL, 2006.
7. *Chemists' Guide to Effective Teaching*; Pienta, N.J.; Cooper, M.M.; Greenbowe, T.J., Eds.; Prentice Hall: Upper Saddle River, NJ, 2005.
8. Moog, R.S.; Creegan, F.J.; Hanson, D.M.; Spencer, J.N.; and Straumanis, A.R. *Metrop. Univ. Int. Forum.* **2006**, *17*, 41- 52.
9. *Education for Judgement;* Christensen, C.R., Garvin, D.A., and Sweet, A., Eds.; Harvard Business School: Boston, MA, 1991.
10. Cooper, M. M. *An Introduction to Small Group Learning,* In *Chemists' Guide to Effective Teaching;* Pienta, N. J.; Cooper, M. M.; Greenbowe, T. J., Eds.; Prentice Hall: Upper Saddle River, NJ, 2005.
11. Abraham, M. R. *Inquiry and the Learning Cycle Approach,* In *Chemists' Guide to Effective Teaching;* Pienta, N. J.; Cooper, M. M.; Greenbowe, T. J., Eds.; Prentice Hall: Upper Saddle River, NJ, 2005.
12. Lawson, A. E. *Science Teaching and the Development of Thinking;* Wadsworth: Belmont, CA, 1995.

13. Farrell, J. J.; Moog, R. S.; Spencer, J. N. *J. Chem. Educ.* **1999**, *76*, 570-574.
14. URL http://pcrest.com. Last accessed, November, 2007.
15. Hanson, D. M.; Apple, D. K. *Process – The Missing Element*, in *Volume IV: What Works, What Matters, What Lasts;* Project Kaleidoscope, Washington, DC, 2004. Available as download at URL http://www.pkal.org/documents/ProcessTheMissingElement.cfm. Last accessed, November, 2007.
16. Lewis, J. E.; Lewis, S. E. *J. Chem. Educ.* **2005**, *82*, 135-139.
17. Hanson, D. M.; Wolfskill, T. *J. Chem. Educ.* **2000**, *77*, 120-129.
18. Dewey, J. *The Child and the Curriculum;* University of Chicago Press: Chicago, IL, 1920.
19. Vygotsky, L.S. *Thought and Language;* The MIT Press: Cambridge, MA, 1986.
20. Ausubel, D. P. *The Psychology of Meaningful Verbal Learning;* Grune and Stratton: New York, 1963.
21. Piaget, J. *The Development of Thought: Equilibrium of Cognitive Structures;* Viking: New York, 1977.
22. Karplus, R.; Their, H. D. *A New Look at Elementary School Science;* Rand McNally: Chicago, IL, 1967.
23. Abraham, M. R.; Pavelich, M. J. *Inquries Into Chemistry, 3rd Edition;* Waveland Press: Prospect Heights, IL, 1999.
24. Slavin, R. E. *Educ. Leader.* **1991**, *48*, 5, 89-91.
25. Slavin, R. E. *Cooperative Learning, Theory, Research, and Practice;* Prentice Hall: Englewood Cliffs, NJ, 1990.
26. Triesman, U. *Coll. Math J.* **1992**, *23*, 5.
27. URL http://www4.ncsu.edu/unity/lockers/users/f/felder/public/. Last accessed, November, 2007.
28. Felder, R. M. *J. Engr. Education* **1995**, *84*, 361-367.
29. Felder, R. M.; Felder, G.N.; Dietz, E. J. *J. Engr. Education* **1998**, *87*, 469-480.
30. Westmeyer, P. *School Science and Mathematics*. **1961**, 317-322.
31. Merrill, R.J. *Sci. Teach.* **1961**, 26-31.
32. Ditzler, M. A.; Ricci. R. W. *J. Chem. Educ.* **1994**, *71*, 685.
33. URL http://madcp.fandm.edu . Last Accessed November, 2007.
34. *New Directions for General Chemistry;* Lloyd, B.W., Ed.; American Chemical Society: Washington, D.C., 1994.
35. Herron, J. D. *The Chemistry Classroom: Formulas for Successful Teaching;* American Chemical Society: Washington, DC, 1996.
36. URL http://www.pkal.org/ . Last Accessed November 2007.
37. Barrows, H. S. *Problem-based Learning: An Approach to Medical Education.* Springer Publishing Co.: New York, 1980.

38. *The Power of Problem-Based Learning: A Practical "How To" for Undergraduate Courses in Any Discipline*; Duch, B. J.; Groh, S. E.; Allen, D. E., Eds.;. Stylus: Sterling, VA, 2001.
39. *Bringing Problem-Based Learning to Higher Education: Theory and Practice*; Wilkerson, L.-A.; Gijselaers, W. H. ,Eds.;. Jossey-Bass: San Francisco, CA, 1996.
40. URL http://www.udel.edu/pbl/, Last Accessed November, 2007.
41. URL http://physics.dickinson.edu/~wp_web/wp_homepage.html. Last accessed, November, 2007.
42. URL http://www.jitt.org . Last accessed, November, 2007.
43. URL http://www.ncsu.edu/PER/scaleup.html. Last accessed, November, 2007.
44. Hestenes, D.; Wells, M.; Swackhammer, G. *The Physics Teacher* **March, 1992**, pp. 141-158.
45. Landis, C. R.; Peace, Jr., G. E.; Scharberg, M. A.; Branz, S.; Spencer, J. N.; Ricci, R. W.; Zumdahl, S. A.; Shaw, D. *J. Chem. Educ.* **1998**, *75*, 741-744.
46. Anthony, S.; Mernitz, H.; Spencer, B.; Gutwill, J.; Kegley, S.; Molnaro, M. *J. Chem. Educ.* **1998**, *75*, 322-324.
47. Gosser, Jr., D. K.; Roth, V. *J. Chem. Educ.* **1998**, *75*, 185-187.
48. Russell, A. A.; Chapman, O. L.; Wegner, P. A. *J. Chem. Educ.* **1998**, *75*, 578-579.
49. Burke, K. A.; Greenbowe, T. J.; Gelder, J. I. *J. Chem. Educ.* **2004**, *81*, 897.

Chapter 2

A Cognitive Model for Learning Chemistry and Solving Problems: Implications for Curriculum Design and Classroom Instruction

David M. Hanson

**Department of Chemistry, Stony Brook University
Stony Brook, NY 11794-3400**

A cognitive model for learning is a representation of the mental processes associated with acquiring new knowledge. In order to be maximally effective, teaching needs to be consistent with a valid model for the learning process. A cognitive model for learning chemistry and solving problems is derived from Johnstone's information processing model, the tenets of how people learn, and research on problem solving. The implications of this model for the development of curriculum materials and instructional strategies are described.

Over 25 years ago Frederick Reif, a physicist at the University of California Berkeley, described a gap that existed between two groups interested in learning: cognitive scientists and educators. He characterized the cognitive scientists as thinking analytically, striving to develop theoretical models, and conducting experiments to test and validate the models, but not being concerned with questions directly relevant to practical educational practice. He characterized educators as being dedicated to their classroom teaching, but approaching their task intuitively and ignoring experimental validation and theoretical models in designing instruction. His conclusion was that "work in education and problem solving could profit substantially if this gap were bridged, if people interested in practical education would build upon the insights and methods of the cognitive scientists, and if educators were to adopt modes of analytical thinking and quality standards of the kind prevalent in other sciences"(*1*).

Much progress in this direction has been made over the last 25 years. The issue has been discussed in articles (*2,3*), books have been written with the intent of bringing research-based knowledge to classroom practice (*4-6*), and a new prestigious Physical Review journal, *Physics Education Research,* has been inaugurated along with an American Chemical Society award for *Achievement in Research for the Teaching and Learning of Chemistry*. The on-going goal, as Reif saw it, is to elucidate how the human learning system works, and use this knowledge to address the complementary issue of how to make it work better. Reif anticipated that this endeavor would lead to several outcomes. Among others, one was the development of models describing mental knowledge structures and the execution of high-level tasks, and another was more effective methods of teaching students (*1*).

In this chapter, a cognitive model for learning chemistry and solving problems is constructed from information in the research literature and implications for the design of instruction and curriculum are identified. The objective is to provide a foundation for the development of more effective methods of teaching students that are based on research in the cognitive sciences.

Cognitive Models for Learning

A cognitive model for learning is a representation of the mental processes associated with acquiring new knowledge. Cognitive models for learning guide the development of curriculum materials, the implementation of teaching strategies, the design of assessment and evaluation instruments, and the research on learning and teaching. Cognitive models can be derived from research in the cognitive sciences, from the continuum of changes that occur as novices develop and become experts in a discipline, and from experimental data that test the validity of proposed models.

For example, the sage-on-the stage model for teaching and learning is illustrated in Figure 1. In this model knowledge is transferred intact from the teacher to the student where it is stored and subsequently accessed and utilized when needed. Being guided by this model, instruction would provide information through lectures and textbooks, and exams would focus on the recall of this information. Research motivated by this model might address the following questions: Does the learner acquire understanding as well as information from the instructor? Can the learner represent the new knowledge in multiple ways and apply it in multiple contexts? Research addressing these questions has been conducted, and the results document that this model fails to describe effective teaching and learning. Teaching and learning is not adequately described as a transfer of knowledge from the instructor to the learner (*6*).

The information processing model, which is illustrated in Figure 2, provides a better picture of the complexities associated with teaching and learning (*7*). In this model, sensory information (visual, auditory, tactile, and olfactory) passes through a perception filter, which is controlled by prior knowledge and experiences stored in long-term memory, before it reaches the working space, where it is processed. Some of the incoming information then is stored in long-term memory and subsequently can be retrieved. Lamba addresses Johnstone's information-processing model more fully in the next chapter.

The information processing model is a significant improvement over the sage-on-the-stage model because it explicitly recognizes that prior knowledge and experiences control what is perceived and can be learned, and that there are

Figure 1. Sage-on-the Stage Model

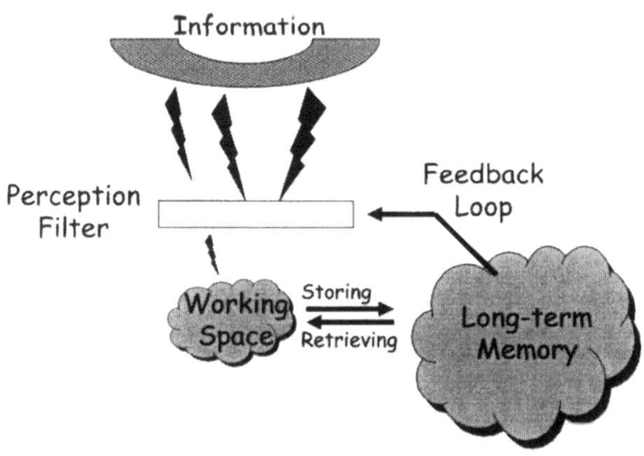

Figure 2. Information Processing Model

two memory components – the working space with a very limited capacity and long-term memory with a much larger capacity. This model has two major implications for instruction (*7*):

1. Prepare students for learning because what they learn and how they learn will depend on and must connect with what they already know and understand and have done successfully before.
2. Do not provide too much material in too short of a time period because the processing rate is limited by both the perception filter and the capacity of the working memory.

The first item connects with the contemporary theory of learning called *constructivism* (*8*). Constructivism means that students come to college with their own experiences and views of the world and try to understand new situations in terms of what they already know and have experienced. The teacher cannot simply dispense knowledge, but rather, needs to help the student in the process of acquiring new knowledge and integrating it with what they already know. Teaching practices that incorporate a role as facilitator or coach therefore are essential.

Research suggests that additional components need to be included in the information processing model in order to have a better representation of how learning occurs. These include an *output processor,* a *perception filter controller,* a r*eflector/analyzer,* and a *librarian.* The output processor accounts for the fact that responses to questions and situations are not always characteristic of what is going on in the working space, and the controller makes it clear that the perception filter accepts conscious control signals from the working space and subconscious control signals directly from long-term memory. The expanded information processing model with these additional components is illustrated in Figure 3.

The Reflector/Analyzer

The *reflector/analyzer* is needed to represent the role of metacognition in the learning process. Metacognition literally means thinking about thinking. It includes self-management, self-regulation, self-reflection, and reflection on learning. Effective learners take charge of their own learning and monitor it (self-management and self-regulation), they think about their performance and how it can be improved (self-assessment), and they reflect on what they have learned and what they do not yet understand (reflection on learning) (*6*). Such metacognition produces an environment for continual improvement and growth and is essential for success and life-long learning.

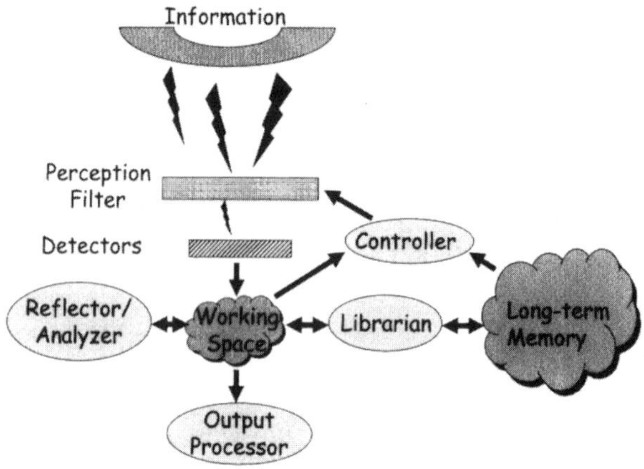

Figure 3. Expanded Information Processing Model

The Librarian

Herbert Simon, a Nobel Laureate and one of the original researchers on human problem solving (*9*) made an important point that underlies constructivism and the role of the *librarian* module in the cognitive model. He wrote, "there is no direct way in which the words pronounced by a teacher can be stored directly as productions available to the student. There must be a conversion of the external language into the internal representation of the student's production system, and neither he nor the teacher knows explicitly what that representation is" (*10*). In artificial intelligence systems, the way information is stored and accessed is known to the programmer, and a compiler program converts the programmer's language into machine language. In human learning, the learner must make the conversion from external representations into an internal representation, and add the necessary connections so the knowledge can be retrieved and used when needed. It is the Librarian module that executes this task in the expanded information processing model.

Only the Librarian knows how knowledge is organized and has been stored in long-term memory and how to retrieve it along with other essential knowledge when needed. Research comparing experts and novices in their approach to solving problems reveals significant differences in the way experts and novices organize their knowledge with respect to both content and structure (*6*).

For this discussion, four knowledge forms are identified: information, procedures, concepts, and contexts. *Information* consists of facts, symbols and their meaning, nomenclature and vocabulary, relationships and equations, and tools. A *procedure* is a process that provides a sequence of steps to produce a result. A *concept* is an idea that connects items in terms of shared characteristics. A *context* is the entirety of some situation.

Novices have a sparse library of discipline content with many equations and relationships memorized, but poorly understood. Experts on the other hand, have a vast library of information, procedures, concepts, and contexts that are totally understood. They visualize this knowledge in multiple representations and can relate these representations to each other, e.g. graphs, diagrams, equations, and words.

The knowledge of novices is not organized, the pieces are disconnected and isolated, and representations are incomplete and not related to each other. The knowledge of experts is arranged and linked hierarchically. There are strong connections between all items: information, procedures, concepts, and contexts. Experts relate different contexts through the underlying concepts while novices analyze the context in terms of surface characteristics (*11*).

The differences between the knowledge content and structures of novices and experts have a significant impact on success in problem solving. The sources of the difficulty novices have compared to experts are summarized in Table I (*6, 11-13*). The key point is that unlike discipline experts, novices do not have their knowledge organized in hierarchical structures in which all the knowledge forms (information, procedures, concepts, and contexts) are strongly interconnected to facilitate access and retrieval. The librarian module in our cognitive model is responsible for organizing and retrieving the knowledge. To improve learning and the ability to apply knowledge in new contexts, classroom instruction therefore needs to be directed at helping the novice's librarian build memory structures that are valid and supercede prior misunderstandings, that are interconnected, and that facilitate access to knowledge when needed. Such instruction is directed at growing the novice librarian into an expert librarian.

Implications for Instruction

In view of the expanded information processing model, teaching needs to help students acquire information efficiently (optimize control of the perception filter), organize it coherently in long term memory (develop an expert librarian), utilize the knowledge effectively (increase the power of the working space), communicate the results meaningfully (enhance the output processor), and reflect on the process to produce continual improvement (turn on the reflector/analyzer).

The POGIL methodology is designed to address these issues. This methodology only is summarized here because it is discussed elsewhere (*14,15*) and in the opening chapter of this book. In a POGIL classroom, students follow a learning cycle of exploration, concept formation, and application. They work together to construct their own understanding and apply it in solving problems in new contexts. Through discussions, they visualize and connect concepts and multiple representations. Lessons end with students reflecting on their progress, assessing their performance, and identifying ways to improve. POGIL activities

Table I. Problem-Solving Process of Novices and Experts

Item	*Novice*	*Expert*
Identification of problem type or characteristics.	Based on surface features.	Based on underlying concepts.
Identification of relevant knowledge items.	Scans through a library of equations to use or manipulate. Has difficulty because information, procedures, concepts, and contexts are not adequately connected.	Conducts an analysis that often includes changing the representation, e.g. drawing diagrams or writing equations. Accomplished quickly because information, procedures, concepts, and contexts are strongly interconnected in large scale functional units.
Addressing particularly challenging problems.	Focuses on one approach and becomes frustrated quickly.	Focuses on concepts and analogies and their relevance to the situation. Looks for multiple approaches and searches for similar contexts in a vast library.
Use of mathematics.	Adds to confusion.	Serves as a powerful tool for analysis, exploration, and problem solving.
Validation of the problem solutions.	Unable to check for inconsistencies because of little information, limited perspectives, and lack of connections within the knowledge structures. Cannot explain why the answer is correct other than that supposedly the correct procedure was used.	Has a vast library of information and multiple approaches to use in checking for inconsistencies. Understands the situation and can explain why the answer is correct.

(*16, 17*) are designed to improve skills in information processing, critical and analytical thinking, problem solving, teamwork, oral and written communication, and metacognition, thereby enhancing the performance of all the components in the cognitive model: the working space, the perception filter controller, the librarian, the output processor, and the reflector/analyzer. While the results of implementing the POGIL methodology generally are positive in terms of student performance and perceptions (see, for example, Chapters 19 and 20 of this volume), additional strategies can be incorporated to support even further gains in problem-solving skills.

Typically problem solving is taught in chemistry lectures, textbooks, solution manuals, and web resources by providing example problems and their solutions. Sometimes a strategy and general steps in a problem-solving process are specified and described. Both of these approaches focus students on the sequence of steps that must be carried out. As result, many students memorize the steps and pattern, and work additional problems that are assigned for practice by pattern matching or by using an algorithm. When the context is changed and the pattern no longer is clear, students do poorly and become frustrated because they have not acquired the knowledge base and skills needed to address the problem in the new context.

Research has documented that to be successful in transferring knowledge and solving problems (*9-11, 13*), students need to be able to:

- acquire the necessary knowledge and integrate it with prior knowledge.
- construct valid and useful long-term hierarchical memory structures of the knowledge that interconnect information, procedures, concepts, and contexts.
- retrieve the interconnected knowledge as needed to solve problems.
- analyze problem situations: identifying the context, assessing what is valid and not valid, what is known or assumed to be known, what needs to be determined or assumed; and what needs to be done to solve the problem.
- recognize the relevance of information, procedures, and concepts in particular contexts.
- synthesize and transfer knowledge to construct problem solutions in terms of the relevant information, procedures, concepts, and contexts.
- reflect on the validity of a problem solution.
- reflect on their performance: identifying possible sources of error based on past performance and experience, and identifying ways to improve in the future.

As reported in the literature (*10, 11, 13, 18-21*), these abilities can be developed by learning activities designed around the following components.

Students work on these activities together and share their insights with each other. The activities have students:

1. identify the important concepts covered in an assignment and the connections between those concepts.
2. reflect on the science content used in solving a problem: what was the key in analyzing the problem and developing a solution, what concepts were needed, why were they needed, what were the steps in the procedure, why were those steps necessary, and how are the concepts related to the procedure?
3. reflect on their problem-solving process in order to improve future performance: what was done, how could the process have been better, and what insights can be identified that will improve future performance?
4. use multiple representations to promote a better understanding of concepts.
5. apply procedures and concepts in multiple contexts and compare and contrast problem situations in order to see similarities and differences and build a library of contexts to draw upon in making analogies. The learner must be specifically aware of the knowledge that is transferable and how it can be transferred (*10*).
6. solve problems in contexts that change incrementally. Some of the knowledge being used in a new task must be identical to some of the knowledge that was learned in a previous task (*10*).
7. classify and categorize what is being learned, identify relationships between the categories, and organize the information and concepts hierarchically rather than in a linear sequence. Concept maps are one approach to this task.
8. make predictions based on their current understanding so they can build on what they already know and encounter deficiencies in their current understanding.
9. develop more than one solution to a problem to force a broader perspective and emphasize that neither the answer nor a particular solution is the important result.
10. employ a specific procedure, such as visualize the problem by drawing a diagram or other representation, analyze the problem in terms of the underlying concepts, develop a plan or strategy employing the necessary equations or relationships to produce an answer, execute the plan, and validate the result. To be successful, such procedures need to be taught, modeled, required, and reinforced (*21*).

11. use general strategies such as means-ends analysis, changing representations, working backwards, and replacing a goal with a subgoal. In means-ends analysis the difference between the current state of the problem and the solution is identified, and action is taken to reduce this difference. Changing the representation (e.g. by converting a word problem to a diagram, table, or set of equations), working backwards, or identifying subgoals or simpler but similar problems often make the path to the solution clear (*13*).
12. receive timely assessment and feedback from both the instructor an their peers on their problem-solving process.

Summary

A cognitive model for learning has been identified. Following Johnstone (*7*), this model identifies the roles of long term memory, perception filters and controller, detectors, a working space, a librarian, an output processor, and a reflector/analyzer in the learning process. The implications of this model for instructional design, strategies, and materials are described.

Telling students concepts and techniques to be used in problem solving and showing them examples generally have proven to be ineffective in growing problem solving skills. Rather students need to analyze examples, practice repeatedly, participate actively, discuss with others, and receive prompt assessment feedback. Teare (*22*) emphasizes that students learn to solve problems by actually solving them. They need repetitive opportunities with continual feedback on their problem-solving process with emphasis on the analysis of the problem and the formation of a plan, not on the solution, and certainly not on the answer. They also need to be given a range of challenges. If problems are too easy, there is little learning, and if problems are too difficult, they become frustrated and give up, and there also is little learning. The level of difficulty depends on the expertise of the learner. Consequently both structured and unstructured problems should be assigned. Structured problems generally have multiple parts, but all the information is given or is readily available. Unstructured problems are more typical of those encountered in the real world, require assumptions, and the concepts that are needed to arrive at a solution are not obvious. To aid in the development of problem solving skills, courses in problem solving typically include the following aspects of problem solving among others: tools for problem representation, models as aids for thinking, identifying personal problem-solving styles and their strengths and liabilities, thinking outside the box and overcoming conceptual blocks, dealing with uncertainty, and problem-solving heuristics (*23*).

References

1. Reif, F. *Theoretical and Educational Concerns with Problem Solving: Bridging the Gaps with Human Cognitive Engineering.* In *Problem Solving and Education: Issues in Teaching and Research;* Tuma, D.T.; Reif, F., Eds.; Lawrence Erlbaum Associates: Hillsdale, NJ, 1980; pp 39-50.
2. Bunce, D.M. and W.R. Robinson. *J. Chem. Educ.*, **1997**, *74(9)*, 1076-1079.
3. Gabel, D. *J. Chem. Educ.,* **1999**, *76(4)*, 548-554.
4. Herron, J.D. *The Chemistry Classroom: Formulas for Successful Teaching*; American Chemical Society: Washington, DC, 1996.
5. *Chemists' Guide to Effective Teaching*; Pienta, N.J.; Cooper, M. M.; and Greenbowe, T. J., Eds.; Pearson Prentice Hall: Upper Saddle River, NJ, 2005.
6. *How People Learn: Brain, Mind, Experience, and School*; Bransford, J.D.; Brown, A.L.; Cocking, R.R., Eds.; National Academy Press: Washington, DC, 2000.
7. Johnstone, A.H. *J. Chem. Educ.*, **1997**, *74*, 262-268.
8. Bodner, G.M. *J. Chem. Educ.*, **1986**, *63*, 873.
9. Newell, A.; Simon, H.A. *Human Problem Solving*; Prentice-Hall: Englewood Cliffs, NJ, 1972.
10. Simon, H.A. *Problem Solving and Education.* In *Problem Solving and Education: Issues in Teaching and Research*; Tuma, D.T.; Reif, F., Eds.; Lawrence Erlbaum Associates: Hillsdale, NJ, 1980; pp 81-96.
11. Bunce, D.M. *Solving Word Problems in Chemistry: Why Do Students Have Difficulties and What Can Be Done to Help?* In *Chemists' Guide to Effective Teaching*; Pienta, N.J.; Cooper, M.M.; and Greenbowe, T.J., Eds.; Pearson Prentice Hall: Upper Saddle River, NJ, 2005.
12. Larkin, J.; McDermott, J.; Simon, D.P.; Simon, H.A. *Science*, **1980**, *208*, 1335-1342.
13. Larkin, J.H. *Teaching Problem Solving in Physics: The Psychological Laboratory and the Practical Classroom.* In *Problem Solving and Education: Issues in Teaching and Research*; Tuma, D.T.; Reif, R., Eds.; Lawrence Erlbaum Associates: Hillsdale, NJ, 1980, pp 111-125.
14. Hanson, D.M., *Instructor's Guide to Process-Oriented Guided-Inquiry Learning.* 2006, Lisle, IL: Pacific Crest.
15. Moog, R.S.; Creegan, F.J.; Hanson, D.M.; Spencer, J.N.; Straumanis, A.R.; Bunce, D.M.; Wolfskill, T. *POGIL: Process-Oriented Guided Inquiry Learning.* In *Chemists' Guide to Effective Teaching,* Vol. 2; Pienta, N.J.; Cooper, M.M.; Greenbowe, T.J., Eds.; Pearson Prentice Hall: Upper Saddle River, NJ, in press.
16. Hanson, D.M., *Foundations of Chemistry - Applying POGIL Principles.* 2006, Lisle, IL: Pacific Crest.

17. See URL http://www.pogil.org/materials/all.php for a complete list of published materials. Last accessed, September, 2007.
18. Heuvelen, A.V. *Am. J. Phys.*, **1991**, *59(10),* 891-897.
19. Reif, F. *J. Chem. Educ.*, **1983**, *60(11*), 948-953.
20. Bielaczyc, K.; Pirolli, P.L.; Brown, A.L. *Cognition and Instruction,* **1995**, *13(2)*, 221-252.
21. Heller, K.; Heller, P. *The Competent Problem Solver*; McGraw-Hill Custom Publishing: New York, NY, 1995.
22. Teare Jr., B.R. *Recapitulation from the Viewpoint of a Teacher.* In *Problem Solving and Education: Issues in Teaching and Research*; Tuma, R.T.; Reif, F., Eds.; Lawrence Erlbaum Associates: Hillsdale, NJ, 1980; pp 161-173.
23. Rubinstein, M.F. *A Decade of Experience in Teaching an Interdisciplinary Problem-Solving Course.* In *Problem Solving and Education: Issues in Teaching and Research*; Tuma, D.T.; Reif, F., Eds.; Lawrence Erlbaum Associates: Hillsdale, NJ, 1980, pp 25-38.

Chapter 3

Information Overload, Rote Memory, and Recipe Following in Chemistry

Ram S. Lamba[1,2]

[1]**Department of Chemistry, University of Puerto Rico at Cayey, 205, Barceló Avenue, Cayey, Puerto Rico 00736**
[2]**Current Address: Chancellor, University of Puerto Rico at Cayey, 205, Barceló Avenue, Cayey, Puerto Rico 00736**

In many institutions of higher education, chemistry is taught in the lecture mode. The laboratory component is conducted to verify principles with known outcomes. As a result, learning chemistry is perceived as memorizing facts and following recipes. This chapter describes the principle theories of teaching and learning using a constructivist approach and how these theories influence the chemistry teaching-learning process. In addition, difficulties in student learning due to information overload and presentation of material in unfamiliar forms are examined. This instructional model emphasizes scientific thinking and associated cognitive skills.

[No] thought, no idea, can possibly be conveyed as an idea from one person to another. When it is told, it is, to the one to whom it is told, another given fact, not an idea. The communication may stimulate the other person to realize the question for himself and to think out a like idea, or it may smother his intellectual interest and suppress his dawning effort at thought. But what he *directly* gets cannot be an idea.

John Dewey (*1*)

Since John Dewey made this statement, not much has changed in the teaching and learning process, particularly in science. It is common to observe in many universities and colleges that new material is presented in the lecture mode in the classroom. Students have been "trained" to accept the ideas of the teachers as facts without understanding them; therefore, they have not developed a habit of questioning in order to understand the material. There is little interaction and interchange between the teachers and the students, or among the students. Furthermore, they miss the relevance of what they are learning and thus, memorize algorithms without conceptual understanding (*2*). Often, instead of questioning what they are being told, students resort to rote memory. Although it may not be an efficient way of learning (*3*), lecturing prevails in the classroom.

The use of the lecture as a teaching tool presupposes that the cognitive abilities, the learning styles and the rate of learning of these students are the same. In other words, it is assumed that they all learn "efficiently" when provided with oral information. As a result, their perception of learning science, in particular chemistry, is to memorize the facts provided in the classroom (*4*).

At many institutions, the percentage of withdrawals, failures and barely passing grades (D) in an introductory chemistry course is unacceptably high. This situation causes many students to change majors from science, therefore reducing the number of students continuing through this pipeline, which, in turn, may have a negative effect on the economic development of the country (*5*). The lecture mode of teaching, and the typical presentation of material as static truths in texts, ignores the fact that science is dynamic, that knowledge is acquired more through interaction than by memorizing facts. In other words, science instruction often is not consistent with the nature and processes of science. The above mode of instruction also denies students the opportunity to work as a team, which is extremely important and necessary in the real world of work (6). At times a small number of students may take the initiative to study in teams, which can provide them with more confidence and allow for the interaction which was missing in the classroom.

The situation in the laboratory component of the course is quite similar. Many laboratories are conducted to verify principles by specific examples. Often, there is no relationship between the lecture and the laboratory components of the course. The laboratory experiments have known results, which are passed on from year to year to incoming students. In order to obtain good grades, the students mimic laboratory processes and try to obtain the expected results without acquiring any conceptual understanding of what they are doing. This is not surprising, since the grading system in many institutions depends basically on the final results (for example, expected percent yield of a product, the numerical value of an ionization constant, etc.).

In contrast, an inductive approach in the laboratory enables students to learn, in addition to manipulative skills, how scientists work and develop reasoning from a specific example to generalized principles. The final outcome is not the only important aspect; what is also important is how students reach this outcome and the reasoning processes which they go through to come to the final result. In this approach, the role of the teacher is to facilitate learning by asking guided questions to stimulate students to reason, rather than giving out information. These questions also help develop students' interactions, thereby focusing on the processes of science as well as the development of conceptual understanding. Here lies the difference between a cook and a chef. A cook follows a recipe without having any knowledge of why certain spices have to be added in a specific order to get a specific taste, whereas the chef knows exactly why the order is important and, in case a specific spice is missing, can make a substitution to achieve the required taste. As science educators, it is more important to prepare "chefs" rather than "cooks".

The information that follows elaborates on the reasons that students resort to rote memory and recipe following. In addition, it addresses the importance of following the Learning Cycle (*7*) in the theoretical and the experimental components of introductory chemistry courses based on data collected for several years showing why this approach (guided inquiry–based learning) is more effective.

Principal Theories of Teaching and Learning

Jean Piaget (*8*) developed a cognitive theory indicating that human beings have mental structures and sensorial mechanisms that interact with the environment. This interaction enables us to assimilate or accommodate information into our existing mental structures. The outcome of this interaction allows us to discriminate between ideas that are important to us and those that are not. In some instances this interaction with the environment confirms or further develops what a person knows by modifying the preexisting cognitive structures, while in other instances it creates confusion or a state of disequilibration between our existing knowledge and that which is being presented to us. Equilibrium is restored by modifying the preexisting cognitive structure until the discrepancy is resolved. This cognitive theory of Piaget, as well as some of the main principles of Lev Semyonowich Vygotsky's theory of learning (*9*), have many implications for teaching and learning at all levels and in all disciplines.

Alex Johnstone indicates that humans "have a filtration system that enables us to ignore a large part of sensory information and focus upon what we consider to matter" (*10*). This aspect of filtration is a positive human characteristic, otherwise individuals would become crazy if required to recall everything seen, heard or lived in a lifetime. What is remembered is what has been experienced and is more meaningful, makes more sense, or is more relevant to the individual.

When confronted with a new situation or new information, each filter system interacts with previous knowledge (held in the long term memory) and helps assimilate or accommodate the new situation. The long term memory provides a mechanism through which the filter system helps to select the information. For any event, observation or instruction to have meaning and to be accommodated in the long term memory, it has to be recognized. Learning is effective if the new knowledge is built upon what is already known, and what can be linked to previous knowledge. This process is called constructivism (*11*). Reactions to any stimuli will be based upon our existing knowledge and preconceived information. Many will not recognize information not previously learned or presented in a radically different manner. This point is well presented by Johnstone (*10*) in Figure 1, where blots that are meaningless to the majority of people are shown. However, when these same blots are presented upside down, all of a sudden it makes sense to us that these meaningless blots signify a dog.

Similar situations happen in many classrooms when information (for example, a chemical formula) is presented to students in a manner different than previously seen or learned. In many introductory texts the formula of acetic acid is presented in different forms: $HC_2H_3O_2$, $C_2H_4O_2$, CH_3CO_2H, H_3CCO_2H among others. For a novice it is not easy to recognize that all of these represent the same species. The meaning of these representations is supplemented by what has already been learned. However, it is assumed that since the information had been provided earlier, the students must have retained it and are able to build upon it. Likewise, chemistry teachers may present very complex demonstrations, such as the iodine-clock reaction, thinking this makes the material more relevant, simpler and easier to understand by the students. The demonstration may have been interesting to observe, and may have even drawn the students' attention. However, how many students really conceptually understand the reaction by a mere demonstration, even if it is followed by an elaborate "cook-book" lab experiment?

Oftentimes two-dimensional structures of DNA or the crystal lattice structure of sodium chloride are drawn in order to convey aspects related to three-dimensional interactions to students. Nevertheless, these two-dimensional structures cannot convey the full concept, even if presented in color or with a Power Point presentation, that students are to learn. Then when a similar question is asked using a different two-dimensional structure, we wonder why students did not acquire the appropriate knowledge of interactions (that can be visualized in three dimensional structures). Most students cannot answer correctly because there was no transfer of the information previously taught; that is to say, it was not linked to students' previously stored knowledge, which was provided two-dimensionally (*10*).

Information Processing Model

Johnstone based his information-processing model, Figure 2, on the work of Baddeley (*12*), who developed a complex model in order to show how the information is manipulated in our brain when it passes through our filter system and is stored in our long term memory. This model shows that in the brain we have a working space where interpretation, rearrangement, comparisons and

Figure 1. Meaningless blots

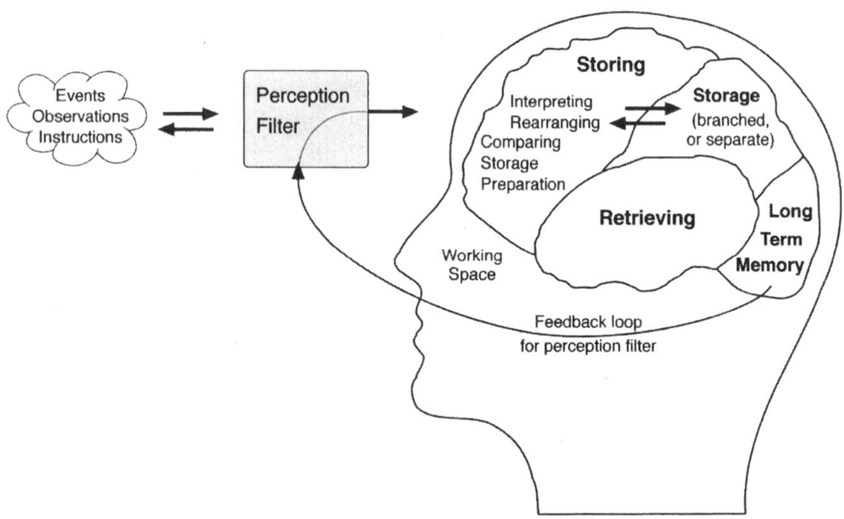

Figure 2. Information processing model

storage of information takes place; and a long-term storage system which stores the previously acquired knowledge. When a person observes any event or listens to information, it is passed through a filter which goes to the working space and then goes on further to the long-term memory. Constant interaction occurs among the working space, the long-term memory and the filter. If the event or information presented is already stored in the long-term memory when presented again in the same form, the filter system helps us recognize it immediately, as illustrated in the inverted dog figure. In the working memory the input is held temporarily for interaction within itself and is stored in the long term memory in order to retrieve information. Short term memory is symbolized by a rapid decay of the input whenever the learner's attention is directed from what is to be remembered. In addition, only a few chunks of information are held in the working memory (*13*).

Similarly, if the student really understood three-dimensional bonding interactions presented in two-dimensional interactions, it would be easy to make a link in the long term memory (preexisting knowledge) and thus understand the new information without difficulty. For many students the information may not be linked with the stored information and therefore is lost during processing in the working memory.

This model also suggests that there is a limit to the number of pieces of information that can be processed in the working space (*14*). When too many concepts are presented in a limited amount of time, it is difficult for students to understand all of them, since there is not much available space to process all this information at one time. When too many different ideas are being presented in one lecture, the mind usually only accepts a limited number of them. Trying to understand everything produces a headache or requires a disconnect from the presentation due to information overload. If the new ideas being presented take more than the working space capacity can handle, then a specific strategy is needed so that the piece of information can be rearranged and managed in order to avoid a "state of unstable overload".

Implications of the Models for Teaching Chemistry

On average, a piece of information (chunks or digits) that an individual can handle and recall from the short term memory is equal to seven plus or minus two digits. Anything above this can cause errors during recall (*15*). If too much information is required for processing, due to the limited working space, very little can be stored.

A similar situation is observed when students are required to manipulate and rearrange information in several forms at the same time. It requires a lot of effort to do several things simultaneously. A simple experiment shown by

Johnstone (*10*) in Table 1 proves the point. The reader can try the experiment by performing the following tasks without writing anything down- everything should be done mentally. The first line in the Table says "Fifteenth of February". Convert this into numbers: 1, 5, 2. Arrange these numbers numerically: 1, 2, 5. This is what is presented in the table. Now, cover the right side of the table for 2-3 seconds and repeat the process. After the third line, most people have difficulty thinking, translating and rearranging the numbers in their heads. This happens due to short term memory overload, which makes one feel uncomfortable, give up or operate below par.

Due to the limitation of working space, strategic learning and working memory have to be used more efficiently. A mechanism that allows a more efficient use of the working space to overcome this limitation is called chunking (*14*). It has been shown that when a child first learns to read every letter occupies a single space in the working memory: for example, m-o-t-h-e-r has six pieces of information. With time this set of letters occupies a single space for MOTHER, instead of six different spaces. Later, when the child starts to read short sentences using the word mother, the whole sentence occupies a single space. This is an example of chunking. Students have similar difficulties until they become familiar with the vocabulary and concepts of chemistry in order to use the working space more efficiently. For us, as chemistry faculty, concepts such as oxidation, reduction, gain of electrons and loss of electrons serve as chunking devices. But it has taken us many years to understand these concepts and be able to use them in our working memory more efficiently. In fact, most of us really acquired full understanding of many of such concepts when we started to teach chemistry.

In introductory chemistry courses, it is taught that the hydroxides (O-H) of first and second group metal ions of the Periodic Table are strong bases. Later, in the Organic Chemistry course students see for the first time a carboxylic acid written as:

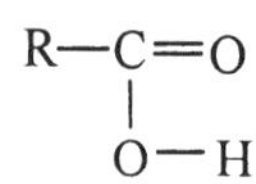

Table I. An Experiment

Fifteenth of February	125
November fifteenth	1115
Fifteenth of February forty five	12455
November fifteenth nineteen forty nine	11114599
December twenty nine nineteen forty one	11122499

This causes an overload in the working memory as the individual atoms are connected in a specific sequence not known to them. Furthermore, the information learned has to change and allow for accommodating the O-H chunk into this new environment. However, in the long run a new chunk, namely "carboxylic acid", uses a single space of information (*16*).

One of the strategies to use chunking devices is learning trends and patterns in the Periodic Table. Similarly, in daily life the chunking device works for any human being to perceive the patterns in terms of the appearance of the full moon every 28 days, going through a specific cycle, and relating the full moon to high tides and the new moon to low tides. Eventually, all this becomes a whole chunk. Chunking usually works when a learner has some recognizable conceptual framework that enables him/her to draw on old or systematize new material. Therefore, for novices, many times the chunking device does not work and thus, they are limited by the available working space until the time they learn the concepts for themselves.

Recipe Following in the Traditional Laboratories

A recent study (*17*) shows the principal difficulties that the students confront in traditional laboratories. Some of these are:

1. Inadequate understanding of the fundamental concepts underlying the experiment.
2. Inability to relate their observations to the theoretical knowledge.
3. Inability to organize their observations so that they are able to dismiss the irrelevant details.
4. Gaps in their knowledge of the subject matter, which not only make it more difficult for them to understand the concept, but may also make them reach a wrong conclusion.

These difficulties are compounded by the way experiments are presented in numerous laboratory manuals available in the market. Many of the manuals cover similar experiments in a manner that encourages students to directly follow instructions without undergoing any thought process (*18*)- that is, by following a recipe. Examining the experiments in many course syllabi, it is evident that every week a new activity requiring the development of new skills and concepts is performed by the students. Although some of the skills acquired need to be applied from one experiment to the other, mostly there is new information, new manipulative skills and ideas to be developed. Thus, it is common that students are expected to learn new concepts every week, without having any chance to use them in a different context or practice them more than once in order to become proficient. Each new experiment introduces and/or confirms new principles, instead of allowing students to develop fewer concepts but with greater depth and understanding. However, if students are allowed the time to apply new skills and ideas in different contexts, and thus make connections, they will not remain first time learners. Most course work requires covering material, leaving little or no time for practice. Therefore, students resort to following instructions (recipes) without understanding the principles behind them. If difficulties are encountered by adults in perfoming new tasks (for example, programming a VCR or a mobile phone), even though adults have more knowledge, maturity and experience than students, why is it expected that students learn new concepts and ideas in the limited amount of time of a laboratory period (2-4 hours)?

Take, for example, when students are asked to perform an iodine-clock reaction. A typical laboratory manual available in the market to study the kinetics of the iodine-clock reaction includes the following instructions:

> Add 10.0 mL of 0.2*M* KI into a 125 mL Erlenmeyer flask. Into the same flask, add exactly 20.0 mL 0.2 *M* KNO_3 from a buret. Next add 1 mL of starch indicator. Finally, add 1.0 mL 0.4 *M* $Na_2S_2O_3$. Label the flask as A and stir the solution thoroughly.
>
> Record the temperature of the solution in flask A and that of a 10.0 mL solution of 0.2 M $(NH_4)_2S_2O_8$. Mix both solutions and record the time of mixing. After mixing, record the exact time that a dark color appears in the solution and remains upon mixing. Immediately add a 1.0 mL aliquot of 0.4 *M* $Na_2S_2O_3$. Mix the solution and wait for the dark color to appear again. When it reappears, record the time and add another aliquot of 0.4 *M* $Na_2S_2O_3$.

When these instructions, which are just a small portion of the whole experiment, are analyzed, over twelve different tasks can be identified including important concepts that the students need to know in order to make sense of their observations. Thus, there should be no surprise that students only follow a recipe and mimic the laboratory movements in order to write their observations in a notebook without engaging in active thinking. If they follow such instructions to perform experiments week after week, what kind of conceptual understanding can be expected of students (*19*)?

Figure 3 shows the tasks for a typical weekly laboratory experiment that a student has to undertake in 2-4 hour period. As can be appreciated from this figure, there is often an information overload in the working memory, thereby causing a state of confusion. Figure 4 illustrates this state of confusion and the student's actions in response to this situation. Under these circumstances, the student follows the recipe, tries to concentrate on a part of the experiment, and

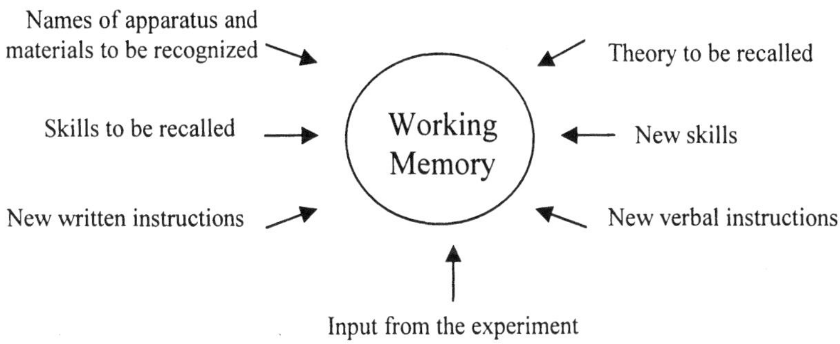

Figure 3. Tasks for a typical weekly laboratory experiment

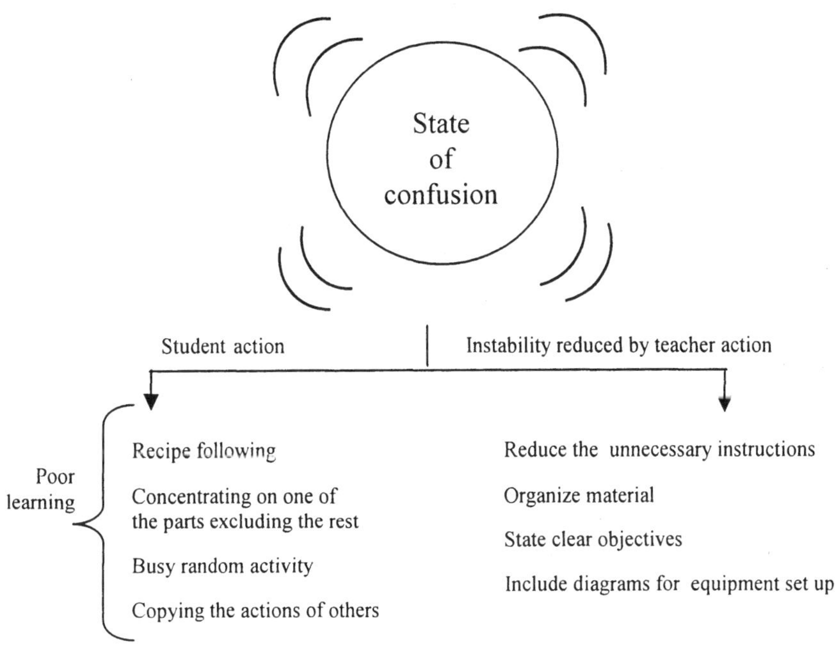

Figure 4. State of confusion, student's actions and the role of a teacher

tries to keep busy by copying the movements of another student who has shown to be good at the laboratory. This, naturally, leads to poor learning.

Figure 4 also shows what actions can be taken by the teacher in order to help reduce the state of confusion and get better results from an experiment. The teacher can avoid giving out unnecessary instructions, organize the material in such a manner that guidance can be provided in short, specific instructions, and include questions which will help students think upon their actions and reactions. Pictures or diagrams which help them understand difficult equipment set-ups could also be used. The specific objectives of the laboratory experiments should be clearly stated without unnecessary verbiage. Furthermore, teachers should try not to give out too many verbal instructions while the students are trying to perform and concentrate on the experiment. All the necessary instructions, hints and precautions should be already provided in the written experiments.

Possible Solutions

In contrast to the traditional mode of giving lectures and recipe- following labs to introduce new material, inquiry-based models show promise in changing the present situation. In this model students learn more when they are actively involved in the process to develop new concepts and integrate them with concepts they already know (*7*). The focus is placed on the learner and the teaching-learning process is based on the following premises:

- The learner is actively involved and participates;
- The learner constructs his or her own knowledge and draws conclusions based on data analysis and discussion of ideas among teachers and peers;
- The teacher places emphasis on questioning as an instructional strategy;
- The learner works in a team to understand concepts and solve problems; and
- The teacher serves as a facilitator in this process.

Finally, the inquiry-based model often focuses on the processes of science as well as the concepts of science as the goal of instruction (*20*). In the case of the guided-inquiry laboratory approach, which follows the learning cycle (see Chapter 1), the experiments have been designed in order for students to use the laboratory experience to develop understanding of key chemical concepts (*20. 21*). The approach encourages scientific thinking in which the emphasis is on "what does the data mean?" rather than "chemically correct" deductions. It also closely reflects what occurs in scientific research.

After the exploration (data gathering) phase, simple questions are asked to stimulate the interpretation and analysis of the data recorded, in order to invent the concept(s) under study. Further reinforcement is provided by assignments which contain questions on the applications of the concepts developed. Both the processes and products of science are important in this approach. The experiments developed using this approach are discussed in more detail in Chapter 16.

In the traditional General Chemistry course, students encounter difficulties integrating chemical concepts that are discussed in the classroom vis a vis the phenomenon that they observe in the laboratory. They generally feel overwhelmed in trying to process the information effectively from the course to the laboratory and viceversa. On the other hand, in the guided inquiry course, the students feel at relative ease in transferring knowledge and information from the course and the laboratory and vice versa.

Conclusions

Examining general chemistry curricula across the nation, it is common that students are expected to learn new concepts every week, presented through the lecture mode. Similarly, each week new laboratory experiments are conducted to develop/verify new principles. Most of the experiments are independent of each other. Throughout the course, for most of the concepts and skills, students remain first time learners and, therefore, resort to memorization and following instructions (recipes).

Research and experience clearly show that it is feasible to part with rote memorization and the recipe-following aspects of the course, giving way to engagement and real understanding of chemical concepts in students. Active involvement of students using hands-on and minds on strategies gives better student outcomes. The changes in students' attitude towards chemistry using guided-inquiry are in accord with the model. This approach has synchronization in its philosophy and style of teaching. The Learning Cycle Approach shows further promise as a learning strategy and its effectiveness is research based. The inquiry-based model emphasizes the processes and products of science, as well as real problem-solving through the experiments.

As faculty members, it is our responsibility to look for models which can be more effective in achieving student learning. Every new model involves a lot more work in the beginning for both instructors and students. However, in the long run, students' achievements and personal satisfaction, as well as attracting more students to become scientists, should be the key factors motivating change.

References

1. Dewey, J. *Democracy and Education: An Introduction to the Philosophy of Education,* Macmillan: New York, **1916/1926**, p.188.
2. Ausubel, D.P.; Novak, J.D.; Hanesian, H. *Educational Psychology: A Cognitive View;* 2nd Ed.; Holt, Rinehart, Winston: New York, 1978. Novak, J.D.; Gowin, D.B. *Learning How to Learn;* Cambridge Univ.: Cambridge, 1984.
3. Ward, R.J.; Bodner, G.M. *J. Chem. Educ.* **1993**, *70*, 198-199. Zoller, U. *J. Chem. Educ.* **1993**, *70*, 195-197. Bodner, G.M. *J. Chem. Educ.* **1992**, *69*, 186-190. Holme, T.A. *J. Chem. Educ.* **1993**, *70*, 933-935.
4. Bodner, G.M. *J. Chem. Educ.* **1992**, *69*, 186-190. King, A. *In Changing College Classrooms: New Teaching and Learning Strategies for an Increasingly Complex World;* Halpern, D.F., Ed.; Jossey Bass: San Francisco, CA, 1994.
5. Hewitt, N.A.; Seymour, E. *Factors Contributing to High Attrition Rates Among Science, Mathematics, and Engineering Undergraduate Majors: A Report to the Sloan Foundation*; University of Colorado, Denver, 1991. NSF; *The State of Academic Science and Engineering;* National Science Foundation Directorate for Science, Technology and International Affairs, Division of Policy, Research and Analysis: Washington, DC, 1990.
6. Astin, A. *What Matters in College;* Jossey-Bass: San Francisco, CA, 1993. Tobias, S. *They're Not Dumb; They're Different: Stalking the Second Tier;* Research Corporation: Tucson, AZ, 1990. Green, K.C. *Scientific American* September/October 1989, p. 475.
7. Lawson, A.E.; Abraham, M.R.; Renner, J.W. *A Theory of Instruction: Using the Learning Cycle to Teach Science Concepts and Thinking Skills;* Monograph Number One; National Association for Research in Science Teaching, Kansas State University: Manhattan, KS, 1989. Lawson, A.E. *Science Teaching and the Development of Thinking;* Wadsworth Publishing Company: Belmont, CA, 1995.
8. Piaget, J. *The Origins of Intelligence in Children;* Norton: New York, 1963. Piaget, J. *Structuralism;* Harper and Row: New York, 1970.
9. Adey, P.; *The Science of Thinking and Science for Thinking: A Description of Cognitive Acceleration Through Science Education;* Innodata Monographs; International Bureau of Education, UNESCO, Geneva, 1999
10. Johnstone, A.H. *J. Chem. Educ.*, **1997**, *74*, 262-268.
11. Bodner, G.M. *J. Chem. Educ.* **1986**, *63*, 873-878. Glasersfeld, E.; *An Introduction to Radical Constructivism;* W.W. Norton & Company: New York, 1984. Driver, R.; Asoko, H.; Leach, J.; Mortimer, E.; Scott, P. *Educational Researcher,* **1994**, *23*, 5.
12. Baddeley, A. *Working Memory;* Oxford Psychology Series Number 11; Oxford University Press: Oxford, UK, 1986.

13. Brunning, R.H.; Schraw, G.J.; Ronning. R.R. *Cognitive Psychology Instruction*; Prentice Hall, Inc.: Englewood Cliffs, NJ, 1995.
14. Johnstone, A.H. *J. Chem. Educ.* **1983**, *60*, 968-971.
15. Miller, G.A. *Scientific American* **1956**, *195*, 42-44.
16. Herron, D.J. *J. Chem. Educ.* **1999**, *76*, 1354-1361.
17. Nakhleh, M. *J. Chem. Educ.* **1994**, *71*, 201.
18. *New Directions for General Chemistry;* Lloyd, B.W., Ed.; Division of Chemical Education, American Chemical Society: Washington, DC, 1994; 34-36.
19. Johnstone, A.H.; Letton, K.M. *Education in Chemistry.* **1991**, *28*, 81-83.
20. Lamba. R.S.; Kerner, N.K. *Inquiry Experiments for General Chemistry: Practical Problems and Applications;* John Wiley and Sons, Inc.: New York, in press.
21. http://www.pogil.org/materials/labs.php

Chapter 4

Advice from a Sage Who Left the Stage: How to Have a Successful POGIL Journey

Andrew R. Bressette

Department of Chemistry, Berry College, Mt. Berry, GA 30149–5016

For those who have been teaching, making the move from a teacher-centered classroom to a student-centered classroom can be both difficult and intimidating. Described below is the rationale for my change and several strategies I implemented that led to my success as an effective POGIL facilitator. Strategies such as thinking about any change as a research project, remaining skeptical about any change, creating buy-in with students and peers, being informed by assessment, and communicating with experts are all discussed as part of my POGIL journey as I left the role of sage on the stage to become a facilitator of active learning.

Having been the sage on the stage for eight years, it was difficult to think of leaving that stage. I've been a successful teacher garnering strong teaching evaluations and developing excellent relationships with my students. I had a diverse student population (chemistry majors, pre-meds, nursing, biology and animal science majors) in my organic chemistry classes, which sometimes made the job difficult, but all in all I was happy.

My teaching style was based on a few simple ideas. I wanted to engage students in class and get them interested in *understanding* organic chemistry rather than simply memorizing it. I asked frequent questions to both probe student understanding and help students learn how to apply what they already knew. I was frustrated by students that simply wanted to memorize their way through organic chemistry and I frequently emphasized the importance of understanding the conceptual underpinnings of topics.

Students often commented that my class was one of the most enjoyable they have had, that they had learned more in my class than any other they had taken, and that I had the ability to take a difficult class and make it accessible to even the slowest students. So, why alter success. After all, I had the answers. I knew how the material related to the bigger picture and what points were key to understanding the main concepts. I had spent much of my adult life learning, understanding, and applying organic chemistry and I was ready to share my passion and knowledge. I was convinced I simply had to present my knowledge in a clear and engaging way and students would understand and apply chemistry as I did.

As I reflected on my teaching, I realized the problem was *I knew* the answers, *I knew* how the material related to the big picture, and *I knew* what points were key to understanding concepts. I made organic seem easy when I presented it, but the students would say when they tried to apply the material on their own, they couldn't do it. I struggled to understand what caused the disconnect and why students didn't seem to get it until I realized I focused on what *I had* done and what *I knew.* I had developed an egocentric view of teaching and learning. To change I realized I needed to develop an understanding of what my students knew, how students were processing knowledge and how to challenge them to think and act more like a practicing scientist. The choice to leave the stage was becoming more obvious.

Treat your Change as a Research Project

As scientists embarking on a research project we have a natural tendency to set goals that we hope to accomplish. Additionally, we consult the literature, create buy-in with our peers and staff, cast a skeptical eye on data and project results, develop strategies to assess prove our results, and continually communicate with our peers and experts in the field. Likewise it is important to think about any classroom change as if it were a research project. I will now summarize how my journey to a student-centered classroom benefited by following this research project model.

I began by thinking about student outcomes I wanted for my classroom. First I decided that I wanted students that were more actively engaged in trying to understand chemistry during class time. Second, I wanted to teach in a way that prevented students from simply writing down what needed to be memorized and moved them towards a true understanding of material. Finally I wanted to give students a chance to practice applying and extrapolating their knowledge to new problems rather than simply regurgitating my examples.

Having set goals for my classroom that focused on student learning outcomes, I reviewed the literature. What successes and failures were already documented? What outcomes had others reported that might help as I move towards achieving my goals? With a critical eye, I began to look at the published pedagogical research (*1 - 6*); a second form of research was to attend workshops with my colleagues. The first workshop was the Multi-Initiative Dissemination (*7 - 9*) (MID) workshop that provided a snapshot of four different innovations in chemical education.

Based upon my reading of the literature and my experiences at the MID workshop, POGIL was the natural choice to achieve the goals I had set for attaining a more engaged classroom. POGIL relied on chemistry content as the driving force in the learning process, using data to guide students toward building an understanding of chemistry, not simply to justify knowledge.

I immediately realized that POGIL would help move students away from memorization and increase student engagement in class. I was equally impressed with the way the POGIL classroom activities were designed. Each classroom activity focused students' attention on the underlying concepts and the interconnectedness of ideas while also building data analysis and problem solving skills. Evidence was presented showing how POGIL had increased student confidence, increased retention (both classroom numbers and content knowledge) and created an environment where students were less intimidated by chemistry.

Be a Skeptic

To meet my classroom goals I was ready to try something new and I was excited POGIL might improve student learning, but I wasn't convinced. As scientists we crave data and analysis but are also trained to be skeptical. POGIL provided data that demonstrated others had positive experiences but what about my students. Would my students accept a new teaching method and have the same success as others? In retrospect, being a skeptic had several benefits that will greatly enhance your experience. First, by being skeptical of doing something different and new, my colleagues were not put on the defense. While it was easy for me to talk about how POGIL might help my students it was equally easy for my departmental colleagues to share my skepticism and even say "I hope this works." Essentially, by being skeptical about the potential benefits of POGIL, I was changing the debate *from how ineffective I thought lecture was to how POGIL might help solve a common problem.* Being skeptical also suggested to my colleagues that I hadn't already jumped to conclusions about what POGIL would do for my students nor was I blindly following the latest fad in educational reform.

Second, being skeptical caused me to really look at what I was doing in a careful and critical way. Did I see evidence that students were able to use data more effectively? Could students apply knowledge to new problems more effectively? Were they more self critical, more engaged, and more analytical? While I never expressed doubt about using POGIL to my students, inside I was full of doubt. In effect, tests, classroom problems, and discussions became opportunities for me to see if the class was really performing adequately without me lecturing. Every day I was looking for evidence of success or failure and sought out ways to gather evidence.

Finally, being a skeptic also meant I had some very practical questions about my change. How could I move to a student-centered classroom? What would I do during the class time? What materials are available to use and will they allow me to teach the same content? Again the literature and some well designed POGIL workshops helped me to overcome these barriers, gave me ideas to engage students, and some behaviors to model. Was I convinced this would work? Not totally, but three weeks after an in-depth POGIL workshop I left the stage and began teaching with student centered classrooms.

Create Buy-in with Peers and Students

With any change, creating buy-in among colleagues and staff is vital to the success of the project. Having the support of my colleagues and administration made a huge difference in my success as a new POGIL practioner. To create this vital support among my colleagues and administration I maintained my skepticism while also sharing the successes I had learned about in the literature. At my institution we were focusing on retention so I made sure my chair, dean, and provost saw the data documenting that POGIL seemed to drastically reduce the number of D, F, and W grades leading to improved retention. The data suggested more students were staying in classes and moving on to the next level. My dean and provost were excited and elated that I was actively seeking solutions to the problems they felt were most crucial for the institution. Not only did this create the buy-in necessary to institute change, it created an environment at my institution where the administration was very interested in my success. After all, if I succeeded in my implementation, they could claim success for their project as well.

To help create support from my colleagues I was clear with them that I was willing to try POGIL but that I had no expectations that they needed to change anything. I found it easier to garner my colleague's support when they perceived it would not require any work from them. Finally, regardless of how convinced I became that POGIL was a better way to achieve student learning, I did not tell my colleagues that what they were doing in their classes was bad or ineffective. Rather I waited for my colleagues to become interested in what I was doing. I shared small stories of success (especially if they were aligned with department or institutional goals) and invited interested colleagues to come visit my classes.

At my institution both the provost and president knew I was doing something very different. They both separately visited my class and were astonished at the level of engagement and learning taking place.

For students it may also be necessary to create some buy-in. My experience has been that the closer students are to high school, the less willing they are to accept change. If my implementation was going to be successful I had to get the students on board as well. At the beginning of the semester I talked with my students and tried to reassure them about what I was doing and how others around the country have been doing the same thing with great success. I shared some of the data with them that especially highlighted that learning in this way would not mess up their plans to get into medical school and that they were likely to gain growth in a variety of process skills.

Additionally, one of the changes I made to my implementation was to ease my students into the transition of a POGIL classroom. While some feel you need to adhere strictly to roles and limit the answering of student questions, initially I answered most questions from anyone and very slowly ramped up my expectations of how I expected them to perform in the classroom. I allowed the students to get comfortable with the group roles and to develop the skills necessary to be moderately successful in them. I also explained frequently why I was doing things in clear and concise terms. For example when telling students I wasn't going to answer their question directly I would tell them I was trying to help them develop skills necessary to approach other problems. I did however, make a point to return to those students after a short time frame to ensure that they had come to resolution with their question. The students appreciated this and quickly learned that I was really on their side when it came to their classroom success. Over time, I gradually increased my expectations requiring more and more of my students.

I also tried to look for meaningful places in the classroom where I could offer affirmation about what they were doing and celebrate student successes. This positive reinforcement that might simply have been to comment that they were asking an excellent question gave students the idea they were doing well and on track. Clearly an important aspect of creating student buy-in is to help students build their own confidence and realize they are also thinking and behaving like a chemist.

Finally, I never told my students they were guinea pigs in an experiment and that I was uncertain about whether POGIL would help them or if I was implementing POGIL correctly. While I may have been apprehensive in my first student-centered classroom I gave the impression to the students that everything was going according to plan. Did everything go smoothly? Of course not, but students never questioned that I was not leading them. As time progressed, I was quickly amazed by what they could accomplish without me telling them anything.

Be Informed by Assessment

In my research lab there are techniques I readily use to see whether I have succeeded in synthesizing desired compounds. Likewise, as I planned my implementation of POGIL, it was important to ask myself how I will know if I have succeeded in meeting my targeted student learning outcomes. While some data, like ACS end of the year exams, may be easy to collect, it is important to design assessment into any implementation plan. Knowing what data I would collect and how it will serve as evidence to support my claim of success (or lack thereof) is important from the beginning of any implementation. Chapter 18 of this volume addresses this issue in great depth.

At my institution I already had a long history of giving the ACS standardized exam at the end my courses so that was an obvious way for me to collect data about student success regarding content knowledge. Because the exam is developed by an external group of peers it prevents the claim that I designed an exam that would benefit my students. In addition to content knowledge I chose to measure growth in learning and process skills by administering the Student Assessment of Learning Gains instrument (SALG) (*10*). Because of my skeptical attitude to see if what I was doing was worth the extra time and effort, and my willingness to administer the survey myself, my colleagues readily agreed to allow me to administer the SALG survey to all students in general and organic chemistry regardless of whether they had experienced a traditional or POGIL classroom.

The SALG survey, unlike traditional end of course institutional evaluations, takes the spotlight off the faculty member and asks students about their perceptions of growth regarding course elements. The first part gathers information about how various elements helped their performance in class such as the text, in class discussions, and working with peers. The second part asks students about their perceived growth in process skills such as their ability to extrapolate knowledge to new problems, communication skills, and analytical reasoning ability. The results of my implementation clearly showed that students in the POGIL sections perceived substantially greater growth than those in traditional classes (*11*). In Chapter 19, a broader study of the use of SALG in organic chemistry is presented.

In today's assessment phobic culture it is important to remember that assessment does not have to be an elaborate project that measures all elements of a course. Rather, the best assessments are small, focused, and centered upon the outcomes you wish to effect in your classroom (*12*).

Communicate with the Experts

With any research project communication is essential. You need to keep abreast of recent developments that may assist you in your project. The research, success, and failures of others may offer new ideas for your project. Equally important are your results, which add to the collective body of research and continue to inform others. Likewise, as I changed to a student centered classroom I found continued communication and collaboration with other experts to be vital to my success as a POGIL practioner.

At my institution there was a second POGIL colleague who implemented change at the same time I did. This was important to my success since I was able to communicate daily about what was going on in my classroom. I was able to share success stories, gain valuable feedback about my ideas, and work with another colleague to overcome barriers. If you are the sole implementer at your institution, it is important to seek out a colleague at another institution in your area to fill this role.

I also took advantage of the expertise present at regional and national ACS meetings. By seeking out POGIL practioners at these meetings I had a chance to share my success, get tips to help the areas I wanted to improve, and learn new ways to improve what I was doing. I also learned much about pedagogical research and received help from others in developing and implementing pedagogical research projects. Just as networking helps further laboratory research, my networking through the Division of Chemical Education at ACS meetings has proven a tremendously valuable part of my professional growth as a POGIL practioner.

One of the strengths of the POGIL method is that it allows students to begin to gain confidence that they are using data and knowledge in an appropriate manner to continue building understanding. Although I was having some success in my classroom I was still uncertain I was implementing POGIL correctly so I decided to invite a senior POGIL practioner to visit my classroom. As this consultant sat in my office before class I had a great epiphany. I had asked this person to come to my class to tell me if I was doing POGIL correctly. I had used the data available to craft an implementation that seemed to work for me but I wanted affirmation that I was doing it the right way. I was seeking the direct affirmation that I now no longer gave to my students!

My revelation from the consultancy was that although there are certain aspects that must be present to have an effective POGIL classroom, I did not have to be just like someone else or do all the things I had observed others doing. *The best practioners of student-centered learning take ideas from many sources and make it fit their personalities, their students, and their needs.* In order for me to succeed as a POGIL practioner and for my students to succeed in a POGIL classroom, I needed to be true to my students and myself. Although I continually try new ideas to help improve student learning I often adapt ideas to make them fit my personality and my classroom.

Five years have elapsed since I stepped of the stage. In that time I've come to know my students better than ever and the class is always fresh. My students are active and engaged and frequently don't want to leave at the end of the fifty minute block. Students have become less afraid of frustration and are willing to try and figure things out on their own. They are faster at making connections, applying material to new problems and the questions they ask me are better, more informed, and often very probing. I'm still a Socrates asking lots of questions, but I've left the stage, become more engaged myself, and have no desire to go back.

References

1. Spencer, J.N. *J. Chem. Educ.* **1999**, *76*, 566-569.
2. Farrell, J.J.; Moog, R.S.; Spencer, J.N. *J. Chem. Educ.* **1999**, *76*, 570-574.
3. Spencer, B.; Gosser, D.K., Jr.; Chapman, O.L. *J. Chem. Educ.* **1999**, *76*, 159-160.
4. Landis, C.R.; Peace, G.E., Jr.; Scharberg, M.A.; Branz, S.; Spencer, J.N.; Ricci, R.W.; Zumdahl, S.A.; Shaw, D. *J. Chem. Educ.* **1998**, *75*, 741-744.
5. Anthony, S.; Mernitz, H.; Spencer, B.; Gutwill, J.; Kegley, S.E.; Molinaro, M. *J. Chem. Educ.* **1998**, *75*, 322-324.
6. Lloyd, B.W.; Spencer, J.N. *J. Chem. Educ.* **1994**, *71*, 206-209.
7. *Multi-Initiative Dissemination Project,* URL http://www.cchem.berkeley.edu/~midp/index.html?main.html&1. Last accessed October, 2007.
8. Burke, K. A.; Greenbowe, T.J.; Lewis, E.L.; Peace, G.E. *J. Chem. Educ.* **2002**, *79*, 699.
9. Burke, K. A.; Greenbowe, T.J.; Gelder, J.I. *J. Chem. Educ.* **2004**, *81*, 897-902.
10. *National Institute for Science Education,* URL http://www.wcer.wisc.edu/archive/cl1. Last accessed October, 2007.
11. Straumanis, A.R.; Bressette, A.R.; Simons, E, unpublished.
12. Suskie, L. *Assessing Student Learning*; Jossey-Bass: San Francisco, CA, 2004.

Chapter 5

Phasing into POGIL

R. Daniel Libby

Department of Chemistry, Moravian College, 1200 Main Street, Bethlehem, PA 18018

Excellent teachers are always assessing their students' success in learning and searching for new creative approaches that will provide their students with the best possible learning opportunities. This chapter suggests a process for helping instructors consider whether or not to change their approaches to teaching, and then presents various options for phasing in POGIL. POGIL can be thought of as divided into three structural elements: learning cycle based activities, use of self-managed class groups, and use of specific roles for group members. The last part of this chapter presents options for partial POGIL implementations that use only one of the structural elements alone or any combination of two of the elements together.

Although we may be continually exploring alternative ideas about teaching, how do we go about deciding if or when we should adopt something new and, if so, how do we implement the adoption? There are two major factors that should bear on the first decision: our level of satisfaction with the current situation in our classes and our knowledge of a new approach that we believe might improve our students' learning. This chapter first deals with whether a change should be considered. Then, assuming that the answer is yes and POGIL is the innovation of choice, a variety of ways to phase POGIL into classes are considered.

Should I Adopt POGIL and What Are my Options?

I have been using POGIL-related approaches in my teaching for 28 years and have been a POGIL workshop facilitator for several years. I am frequently asked questions about why or how one might adopt the POGIL approach to teaching. These questions usually focus on whether to use commercially available activities or write personal ones and how often and in what context should POGIL group activities be used. I believe that my experience with phasing in my current POGIL implementation over many years provides additional options to those considering the POGIL approach.

The Decision Process

Before exploring approaches to POGIL implementations, one should revisit the question, "Why should I consider changing the way I teach?" This question can be considered through a series of steps.

1. Am I absolutely satisfied with my current student outcomes? If the answer is yes, go no further.
2. Have I seen alternative teaching models that I believe might improve my student outcomes? If not, begin to explore alternatives that are currently being used.
3. What is my comfort level with potential changes I have considered? Separate aspects that present little discomfort from those perceived as more problematic, major barriers to implementation.
4. What changes are reasonable right now to begin moving toward a new teaching model? It is not necessary to change everything at once.
5. What support is available to assist me with my change? Take advantage of all available practitioners and programs.

For the purpose of this chapter, let us assume that your answer to step one is NO and in answering step two you have discovered the POGIL approach and have done some background reading or attended a workshop. You also believe that POGIL may improve your students' learning experience in one or more of your courses. So, you can see potential advantages in using POGIL, but are unsure whether it will work with your students and whether you can effectively facilitate group activities in your course.

Currently, there is great diversity in POGIL implementations in terms of the sources of course materials and how often group activities are used. There are commercially available activities for several college courses (*1- 7*), but several implementers write their own activities even for courses that have available materials. The other primary variables in these implementations are how often and in what context group activities are used. POGIL activities may be used with in-class groups for essentially all class meetings, they may be used in varying frequencies in class as part of a predominantly lecture course, in recitation sections that support a standard lecture course or in special discussion sections replacing one lecture per week. Each of these approaches is being used in classes varying from about 10-250. The fourth implementation, Peer-led Guided Inquiry (PLGI), employs upper-level undergraduates as facilitators in a large course (*8*).

With the above models available, the doubts of a potential new implementer usually relate to lack of confidence in facilitating group activities and unfavorable classroom design. These concerns often seem insurmountable. However, the current models outlined above for adoption are more limited than they need to be. Below, a broader analysis of the aspects of POGIL is presented along with a number of alternative ways of moving toward POGIL.

Structural Elements of POGIL

One way to deal with step three, is to analyze the POGIL approach in terms of the structural elements of its implementation.

POGIL is based on research indicating that teaching by telling does not work for most students (*9*), that students who are part of an interactive learning community are more likely to be successful (*10*), and that knowledge is personal; students enjoy themselves more and develop greater ownership over the material when they are given an opportunity to construct their own understanding (*8*). Personal construction of knowledge requires several process skills. So, POGIL is designed to use disciplinary content to teach both content knowledge and the process skills necessary to work in the discipline (*11*).

A complete POGIL implementation can be divided into three structural elements:

a. Activities that are generally based on the Learning Cycle, and that are designed for use in self-managed groups to help students "discover" the course content – Constructivist approach developing both content know-ledge and some process skills.
b. Group work – Introduces the social aspect of learning – basis for an inter-active learning community.
c. Group Roles – Specially emphasize development of process skills.

Abraham (*12*) has recently provided an excellent description of the Learning Cycle. More information about effective management of the group-learning aspect of POGIL can be found in Chapter 7.

An Example of Phase-in of POGIL by Structural Element

Thinking about POGIL implementations in terms of these three elements provides a faculty member with a framework (and options) for phasing in a new POGIL implementation. In fact, my current POGIL implementation progressed through items a., b. and c. sequentially over about 20 years. The particular sequence of developments was related to my constant search for better ways to teach (Step 2 of the decision making process).

Implementation of Element a., Learning Cycle Activities

In graduate school (Penn State University, 1969-1974), I discovered the mechanistic approach to understanding organic chemistry. I had always aspired to a small college teaching/undergraduate research career. When I started to see the mechanistic relationships among chemical concepts, I was truly inspired to go out and "spread the good news" to my students. No one had told me about this when I was an undergraduate, so I firmly believed that my students would be freed from the yoke of memorization and would rapidly come to understand how chemistry "works". Imagine my distress when I went off to a visiting assistant professor position at Oberlin College and told my students the "good news": I explained, in detail, how it worked and they didn't understand. No matter how clear my explanations, some students just couldn't succeed. However, I did notice a pattern of student development; they would struggle for a period of time and then there would be a significant increase in their understanding over a very short time. Once students made this jump to higher exam grades, very seldom did they drop back to their earlier lower grades. So this change seemed to be true enlightenment rather than a connection with the topics on one exam. These experiences with student enlightenment motivated me to find a better way to facilitate upward transitions in student understanding. So after my first year of teaching I had my answer to step one and I began looking for an alternative to the lecture model for teaching organic chemistry (Step 2).

After three years of visiting positions at Oberlin and Kenyon Colleges, I obtained a continuing assistant professor position at Skidmore College. This provided a third group of very good students and the pattern of student transitions continued. In the fall of 1978, I attended an NSF Chautauqua Short course, "Development of Reasoning and College Science Teaching" which was offered at Hampshire College. The facilitators were Bob Fuller, a physicist, and Mel Thornton, a mathematician, both from Project ADAPT at the University of Nebraska-Lincoln. This experience provided me with Piaget's constructivist model for learning and the learning cycle process for assisting students in learning (*13,14*). With this minimal experience, I converted my introductory organic course from lectures to learning cycles (*15,16*), starting with the spring semester of 1979 (Steps 3 & 4).

It should be noted that this first implementation of the learning cycle did not include class groups. I used individual activities, hand written on spirit masters (ditto sheets), to stimulate class discussions that led the class members to work together to explore the data (Exploration Phase) and eventually discover the concept (Concept Invention Phase). Students were expected to work on the activities outside of class and come to class with their thoughts on the questions in the activities. Although I encouraged students to work together outside of class, I had no good model for in-class group work. I acted as a facilitator of the whole class discussion to assist students in exploring data and reaching consensus. The Skidmore students, who as a whole had not been thrilled by organic lectures, took to the learning cycles very quickly and enthusiastically. The quality of their interest in organic chemistry and their confidence in their understanding improved greatly. We had been using an ACS standard organic exam as the second semester final. Over my three years at Skidmore there was no statistically significant difference in the ACS organic exam grades. So, as I moved from 2 semesters of lecture to 1 semester of lecture and 1 semester of learning cycle and finally to 2 semesters of learning cycle, the students' abilities to answer the ACS multiple choice questions were not harmed. However, as indicated above, their apparent attitudes improved greatly. I viewed these results as positive.

At Skidmore my organic classes ranged from 15 to 25 students in a single level classroom with movable chairs, so most of the students could be induced to become involved in the class discussion. However, in 1982 I moved to Barnard College where the organic chemistry class was taught in a tiered lecture hall and had 110 students. My learning cycle approach required some adaptation, but after a short adjustment period, the students took to it. It worked quite well even though only a relatively small proportion of the students could actually participate in the all-class discussions. The key to success seemed to be assuring that the discoveries of concepts came from student discussion rather than from the facilitator. I had internal support from my department chair, Bernice Segal, and continued support from the Piaget community through 2 conferences in 1982 and 1985 (*17,18*). These conferences had presentations on learning theory as well as learning cycle implementations in courses across the college curriculum. The interactions at these conferences provided me with many ideas of ways to improve my class activities and support of a group who were committed to active learning approaches (Steps 2 and 5). Unfortunately, no new conferences were held by this group after 1985 and for various reasons, I lost contact with the Piaget community.

Implementation of Element b., Self-Managed Class Groups

The key connection that led to my next major teaching development, implementation of class groups, was meeting Rick Moog at a Gordon Conference on Innovations in Chemical Education in 1993. At that time Moog and two Franklin and Marshall College colleagues, Jim Spencer and John Farrell, had developed what they call a guided inquiry approach for teaching general chemistry. That meeting led to an invitation to join the Middle Atlantic Discovery Chemistry Project (MADCP) consortium. Beside the moral support I received with this exciting group of small college chemistry faculty (Step 5), I was exposed to the small group model that they used at Franklin and Marshall College in their general chemistry course (Step 2).

I had been aware of the value of having students work in small groups, but had not encountered a workable model. Although my teaching style uses considerable all-class discussion, I found that the self-managed small group classroom structure provided a base for helping students develop hypotheses that could then be shared and tested through all class discussion (Steps 3 and 4). So, after seventeen years of working in isolation on my teaching, I once again had a circle of creative and innovative colleagues to provide new ideas and an institution, Moravian College, which truly valued innovative teaching. The fourteen years since my discovery of the MADCP group have been the most stimulating and exciting in my teaching career (Steps 2 and 5).

Implementation of Element c., Special Emphasis on Group Roles

In my subsequent work with the POGIL Project, I have come to appreciate the value of explicitly teaching process skills. When I first began developing learning cycles for my courses, I focused on what I wanted students to be able to do when they completed the course (*16*). My activities were designed to lead students through the processes I use in understanding organic chemistry. So, the analytical process skills were imbedded in the activities. However, in working with my POGIL colleagues, I have come to appreciate how individual roles in self-managed small groups can be used to help students develop specific process skills as they are exploring the course content. Once again collaboration with creative colleagues has help me to continue to develop as a teacher (Steps 2-5).

Lessons Learned

The chronology above describes the three-phase process I used to fully implement POGIL in my organic chemistry class. The current state of my course organization and my activities are available on my website (*19*). In creating my course I have come to appreciate some critical factors in devising a course structure and choosing a teaching method.

- Teach your personal logic for understanding and applying the content rather than using a textbook sequence that is not comfortable (*15*).
- Develop specific learning objectives for your course in terms of what you would like students to be able to DO when they complete your course (*16*).
- Use only teaching approaches that make sense to you.

Types of Partial Implementations

In addition, my experience suggests that there is more than one sequence for adopting the structural elements of POGIL. I implemented one element at a time because of the methods available at the time and my comfort with their use. I adopted the elements one at a time in a particular order: learning cycles, class groups and then emphasis on group roles. However, other implementation sequences may be more logical for others.

Single POGIL Element Implementations

- Learning cycle activities can be used without class groups as a way to move from lecture to active learning in essentially any classroom structure.
- Group work can be used alone to have students discuss lecture material or to solve application problems in recitation sections. Research indicates that learning is a social activity and results from the Peer-Led Team Learning project support the value of group work (*20*).

Two POGIL Element Implementations

- Groups with specific roles can be used as described above with lecture or recitation materials. Group roles can emphasize the process skills critical in "doing" science.
- Learning cycle activities could be used with class groups that have no specific roles to gain the active learning and social interaction without concerns about enforcing specific group role requirements.

As faculty members gain experience with implementing POGIL structural elements they may find that they become satisfied with the element(s) implemented. Alternatively implementation of one element might provide incentive to implement a second or third element. On the other hand, a faculty member might find that a partial POGIL implementation seems optimum for a given situation.

How do I Decide Among Partial Adoption Strategies?

The key to deciding on an implementation is an assessment of perceived barriers to POGIL implementation. Commonly perceived barriers include:

1. Organized group work would be difficult due to classroom design (tiered lecture hall) or complexity of maintaining group organization and assigning grades.
2. Concern that coverage of material will suffer if lecture time is committed to group work.
3. Lack of confidence in facilitating group work with POGIL activities.
4. Lack of confidence that POGIL activities will provide sufficient learning efficiently enough.

For barriers 1 or 3, faculty members could use POGIL activities to stimulate class discussion of a few topics. Appropriate commercial POGIL activities that are comfortable for the faculty member could be chosen (*1-7*) or notes from a few lectures could be converted into POGIL activities (*16*). I suggest that each activity be given out to students on the class day before it is to be used in class and students be instructed to work though the activity either alone or with some classmates before coming to class. In class the instructor can solicit a few responses to each question in the activity. It is important to accept all responses without expressing judgment as to their quality. When it is clear that there is general agreement on a response, simply move on to the next question. When there is a lack of agreement, ask students to provide supporting arguments for each point of view and encourage students to explore the relative merits of each proposal to select or create the best option. If necessary, ask questions that will lead students to discover the weakness in poor suggestions and/or the strengths of better ones. If at all possible, manage the discussion to allow the students to reach the desired conclusion rather than presenting the "correct" answer. When students ask if their conclusions are correct, point out that if they work and everyone agrees then they must be fine. Always seek to build student confidence in their ability to draw reasonable conclusions.

The above implementation could also use informal student groups in class. At the beginning of class, the students could be instructed to compare their responses with one or more classmates before the class discussion begins. This will give the instructor a sense of the real level of difficulty students might experience with formal group work when classroom design is the major perceived problem.

For barriers 2, 3 or 4, POGIL activities could be used occasionally in recitation sections (fourth hour often associated with large lecture classes). Usually these sections are relatively small and should lend themselves to using groups with or without assigned roles. This type of implementation will allow the faculty member to see how effective the activities are at enabling student learning and give the instructor experience with facilitating group work in a less complex situation than with the whole class.

For barrier 4, the recitation sections could use typical application problems, but have students work in organized groups with specific roles. This is similar to the Peer-Led Team Learning (*14*) approach, but in this case the instructor acts as the facilitator without the peer leaders. I suggest that once the faculty member is comfortable with facilitating group work, a few POGIL activities be substituted for the application problems. This would allow the instructor to assess the relative merits of POGIL activities *vs.* application problems.

In addition to these ideas, several chapters in this volume provide further suggestions and information. Chapter 6 deals specifically with issues related to large classes and Chapter 7 discusses effective classroom facilitation strategies. Chapters 19 and 20 provide data concerning the performance of students in POGIL classrooms.

Reflections

My response to step one of the decision process will always be NO regardless of how well my students learn. Although I am convinced that the POGIL approach is my best mode of teaching, I can always see that it could work better. The kind of interactions I have with my students and their enthusiasm for learning keep me searching for ways to make their experience better (Step 2).

As described above, I have occasionally been able to make significant transformational changes in my teaching, but each of these was inspired by an experience with one or more colleagues or mentors. So although technological applications for teaching have developed immensely over the last twenty-eight years, they have had minimal effect on the quality of my teaching. Certainly the documents I produce are of higher quality than the handwritten ditto sheets I used at Skidmore College in the seventies, but the quality of my teaching has been most affected by developments related to my experiences with colleagues described above. These experiences provided answers to steps 2 and 5. All of my major transitions in teaching involved specific individuals, who were willing to share their ideas and accomplishments. When I can see a change that I believe will improve my students' learning (step 2), my personal needs force me to very rapidly find answers to steps 3 and 4.

Consequently I highly value my current involvement with the POGIL Project. The Project is a continually renewing organization designed to assist an ever-increasing group of faculty members to reach their potential as teachers. Each new person who joins the project brings a slightly different way of thinking about the teaching process. We are all looking for better ways to help students learn and that makes all the difference.

References

1. Moog, R. S.; Farrell, J. J. *Chemistry: A Guided Inquiry, 3rd Edition;* John Wiley & Sons: New York, 2006.
2. Hanson, D., *Foundations of Chemistry: Applying POGIL Principles*; Pacific Crest: Lisle, IL, 2006.
3. Garoutte, M. P. *General, Organic, and Biological Chemistry: A Guided Inquiry*; John Wiley & Sons: New York, 2006.
4. Straumanis, A. *Organic Chemistry: A Guided Inquiry*; Houghton Mifflin: Boston, MA, 2004.
5. Spencer, J. N.; Moog, R. S.; Farrell, J. J. *Physical Chemistry: A Guided Inquiry Thermodynamics*; Houghton Mifflin: Boston, MA, 2004.
6. Moog, R. S., Spencer, J. N. and Farrell, J. J. *Physical Chemistry: A Guided Inquiry Atoms, Molecules, and Spectroscopy*; Houghton Mifflin: Boston, MA, 2004.
7. March, J.; Caswell, K.; Lewis, J. *Introductory Chemistry Modules: A Guided Inquiry Approach, Preliminary Edition;* Houghton Mifflin: Boston, MA, 2008.
8. Lewis, S. E.; Lewis, J. E. *J. Chem. Educ.* **2005**, *82*, 135-139.
9. Johnstone, A.H. *J. Chem. Educ.,* **1997**, *74*, 262-268.
10. Elmore, R.F. In *Education for Judgment: The Artistry of Discussion Leadership;* Christensen, C.R.; Garvin, D.A.; Sweet, A., Eds.; Harvard Business School: Boston, MA, 1991.
11. Farrell, J.; Moog, R. S.; Spencer, J. N. *J. Chem. Educ.*, **1999**, *76*, 570.
12. Abraham, M.R. In *Chemists' Guide to Effective Teaching;* Pienta, N.J.; Cooper, M.M.; Greenbowe, T.J., Eds.; Prentice Hall: Upper Saddle River, NJ, 2005.
13. Piaget, J. *J. Res. Sci. Teach.* **1964**, *2*, 176-187.

14. Karplus, R.; Thier, H. D. *A New Look at Elementary School Science,* Rand McNally: Chicago, IL, 1967.
15. Libby, R. D. *J. Chem. Educ.* **1991**, *68*, 634-637.
16. Libby, R. D. *J. Chem. Educ.* **1995**, *72*, 626-631.
17. Libby, R. D., "Organic Chemistry: A Learning Cycle Approach", in *Proceedings of Reasoning, Piaget and Higher Education: 3rd Annual Conference*, Denver, CO, 1982.
18. Libby, R. D., "Reasoning Through Organic Chemistry (or any other Course)", in *Proceedings of the 4th Conference on Reasoning and Higher Education*, Boise, ID, 1985.
19. URL http://www.chem.moravian.edu/~rdlibby. Last accessed October, 2007.
20. Wamser, C. C. *J. Chem. Educ.* **2006**, *83*, 1562-1566.
21. Gosser, Jr., D. K.; Roth, V. *J. Chem. Educ.* **1998**, *75*, 185-187.

Chapter 6

POGIL Implementation in Large Classes: Strategies for Planning, Teaching, and Management

Ellen J. Yezierski[1], Christopher F. Bauer[2], Sally S. Hunnicutt[3], David M. Hanson[4], Katie E. Amaral[5], and James P. Schneider[6]

[1]Department of Chemistry, Grand Valley State University, 1 Campus Drive, Allendale, MI 49401
[2]Department of Chemistry, University of New Hampshire, Durham, NH 03824
[3]Department of Chemistry, Virginia Commonwealth University, 1001 West Main Street, Richmond, VA 23284
[4]Department of Chemistry, Stony Brook University, Stony Brook, NY 11794
[5]Division of Science, The Pennsylvania State University at Berks, Reading, PA 19610
[6]Portland Community College, Portland, OR 97216

POGIL implementation strategies organized around issues associated with the large classroom lecture have been compiled by instructors at various institutions and are presented. Issues such as physical space, group and individual accountability, establishing groups, and stakeholder buy-in are addressed.

Chemistry instruction in a large lecture setting is a fixture in higher education at many institutions. Economics and demographics suggest that colleges and universities will never have enough financial resources or competent and experienced instructors to make 20 to 30 students in a classroom a uniformly achievable goal. Technology-enhanced courses (e.g. distance learning), despite much early enthusiasm, have not demonstrated the ability to reduce costs per student nor expand the student-faculty ratio (*1-7*).

Much has been written about the challenges of teaching and learning in large class settings (*8-12*). In any class, the goal is to engage students in higher-level thinking, to develop students' understanding, and to lead students toward applying knowledge in new situations. This goal can be more difficult to achieve in large classes, where each student is one among many. Large classes create an atmosphere in which perceived personal accountability, intellectual engagement, and motivation are low (*12, 13*). Another challenge is that the accepted mode for large classes is lecture presentation (*13*). The traditional lecture focuses on transmission of information, which suggests to students that learning means making a record of (note-taking) and repeating (memorizing) what the instructor says (*14*). Moreover, lecture presentation (*15*), in its most passive forms, is not particularly effective in helping students attend to and later recall presented material. Exclusive use of lecture may select preferentially for students who already possess mature metacognitive ability and leave other students behind.

POGIL is an instructional model that has been used, primarily in small classes, to achieve the goal of improving student learning. This presentation is intended to demonstrate how it is possible to successfully implement POGIL in a large class by analyzing how perceived impediments to implementation of POGIL can be managed. The definition of "large" for our purpose is a class size above about 40 students; POGIL has been implemented in class sizes up to 300 students. Chapter 12 in this volume provides examples of large classroom implementations. We will not consider here implementation of POGIL in recitation subsections associated with a large class. This situation is parallel to a small class implementation, and there are established examples to use for guidance (*16*).

The following sections discuss POGIL implementation strategies organized around issues associated with the large classroom lecture. These issues are: class organization, physical space, group and individual accountability, distribution and collection of materials, classroom civility, establishing groups, stakeholder buy-in, diversity, and recruitment of student facilitators. The strategies discussed are by no means exhaustive, but are rather a compilation of contributions from experienced instructors from different types of institutions. Readers should evaluate the strategies and approaches according to the needs of their students and institutions.

Class Organization

Clearly, the main challenge to implementing POGIL in a large class is the large number of students. One experienced POGIL instructor alone can manage about 10-15 groups (of three to four members each). Larger classes generally require multiple facilitators (typically graduate or undergraduate students) responsible for three to five student groups. Hence, the classroom contains a primary instructor and assistants who provide direct support to their student groups and communicate with the course instructor.

Pacing of student work is also challenging in a large class. If too much contiguous time is devoted to POGIL activities, facilitation becomes difficult as student groups become more out of phase with respect to their progress on the activity. Several approaches, listed below, help solve this problem.

- Use *personal response systems* ('clickers') to regulate the pace of groups. Posing clicker questions creates deadlines within the activity and keeps students on task. The questions also serve as a formative assessment that allows the instructor to check in with a large number of students quickly.
- *Write shorter large-class activities* designed to take about 20 minutes and concentrate on one or two concepts. The activity should be consistent with the POGIL format and employ the learning cycle.
- *Align the complexity of large-class activities with the allotted time.* An activity that is too short can undermine the development of process skills and cause students to stray off task soon after the activity is assigned. One that is too long can frustrate students if they do not have adequate time to explore and explain the model. If authoring activities, a POGIL writing guide (*17*) is a useful reference.

Working with the Physical Space

Many lecture halls have long rows of closely spaced, fixed, tiered seats. It may be difficult for students to arrange themselves physically in groups of three or four. However, three or four students sitting in a row can form a team by leaning into one another to discuss ideas and look at notes or a handout. Teams of four may also be achieved with two students in one row sitting behind two students in a different row. Setting up a seating chart in advance and posting it electronically saves class time and helps to quickly seat a large class in an effective arrangement. The seating chart is also a convenient way to assign group roles (see *Establishing Groups*).

Often facilitators cannot move easily through a large, crowded lecture hall. They may have difficulty communicating with some groups. Communication between teams and the instructor and facilitators can be assisted by one or more of the following:

- Leave entire rows empty (if possible), so that facilitators can move among the students;
- Scatter student facilitators around the room;

- Create intergroup networks that require teams to consult with other teams instead of, or before, direct contact with facilitators;
- Require facilitators to pre-screen team questions;
- Use electronic methods (cell phones, 'clickers', PC tablets) to send questions to the instructor or facilitator;
- Hold up white boards (or similar) with questions posted on them.

Finally, it may be more challenging to establish a norm of adherence to group roles in the large class. Because roles help develop interdependence (part of the PO in POGIL), special attention should be given to efficiently establishing roles and to preparing assistant facilitators to reinforce them. The following may assist instructors in reinforcing group roles in a large classroom.

- Use cards, colored paper, headbands, or some other distinct marker to tag managers or the other team roles.
- Hold only the manager responsible for communicating with facilitators, minimizing the number of students each facilitator needs to communicate with during class.
- Place managers in the most accessible locations in the lecture hall, such as the ends of rows, giving the entire team better access to facilitators.
- Rotate teams and individual members from inaccessible to accessible seats periodically during the semester.

Establishing Groups

Assigning groups is critical to the success of the POGIL classroom (*17*), but different approaches are required for large versus small classes. A group of three to four students is ideal, but groups of five to seven may be desirable to adjust for absences in a large class. This could be managed by assigning a joint 6-7 member group which operates as two separate groups if everyone is present and a single smaller group if there are several absences. After the add-drop period has ended, groups should be assigned to specific seats using a seating chart (see *Physical Space*) and to avoid too much shuffling of students between groups. If the course has recitation/discussion meetings, groups can be assigned in the first sessions. Alternatively, group lists may be posted electronically. During the first week, students can work with others sitting near them, and formal groups can begin working the second week. Randomly assigned groups may work well and be easier to assign, although Johnson, Johnson, and Smith (*18*) recommend establishing groups to maximize diversity based on achievement, gender, ethnicity, and interests. WebCT has the capability of randomly assigning groups of a specified size as well as by other criteria (e.g. results of a learning styles inventory assigned by the instructor). More negatives are associated with changing groups in a large class situation than in small ones. It takes time to re-form groups, students lose contact with the few people they know, and the number of poorly functioning groups may not be reduced because the instructor cannot know all the students.

Maintaining Individual Accountability

Individual accountability can be envisioned on two levels. The first level is individual accountability in terms of course outcomes. This is the normal expectation that there will be assignments or examinations in the course that require individual effort and evaluation. The instructor must emphasize that group work in class does not replace the need for students to do their own independent thinking, via personal class notes, homework, and reading.

The second level is individual accountability to the group effort. Each person must be seen by the group as contributing to the effort in productive ways. This is a critical component of effective cooperative learning because it minimizes behaviors described as social loafing, in which some students working on a group product let other students do all the work (*18*). Assigning student roles within groups, as mentioned previously, is one way to promote equitable contributions. Another approach is based on the design of materials. Materials could first involve students working in parallel on similar but not identical tasks, then lead to comparisons among group members to check results or combine information to reach a new inference or conclusion. Other successful strategies have been used:

- Include a collaborative problem solving section on quizzes or exams. This typically comprises a much smaller part of the exam than the individual component.
- Assign group quizzes using scratch-off bubble sheets, such as IF-AT forms from Epstein Educational Enterprises, with each group getting one quiz and one form.
- Award a group bonus for individual performance on exams or quizzes. If all the members of a group score above a certain cut-off, the entire team is rewarded with a small bonus. This encourages students to work together and assist each other throughout the course. The group bonus is only appropriate if grades are not curved.
- Use the "team" functionality of clickers (if implemented on your system), or designate one student's clicker as the recorder for the entire group. Students come to a consensus on the answer the entire group wishes to report.
- Groups use the final minutes of one class a week to comment on the performance of the group and contributions of group members by

identifying two or three strengths and two or three things needing improvement.

Managing Classroom Materials

Distribution of materials can be especially time-consuming in a large class. Electronic distribution of new activities before class addresses this difficulty. Returning materials is also difficult, but can be managed using group folders. If groups are assigned (as discussed above), each group is given a folder where their group quizzes, completed activities, reflections, or evaluations are placed. Undergraduate or graduate student facilitators can help create and sort folders, assign credit for completion, or summarize common questions for the instructor. Finally, the instructor may choose to collect materials from a subset of groups or individuals, changing the selection each class.

In addition, using clickers throughout class can provide formative assessment for students and feedback to the instructor. The data provided by clickers (i.e., histograms of answers) are not sensitive to individual groups or students, but give an aggregate picture of content mastery. This provides real-time feedback to the instructor and to the class about where understanding may vary. Collecting work for review and summary will take more time and be less timely.

Classroom Civility

Classroom civility must be addressed in all large classes; however, some characteristics of POGIL classes require specific strategies to keep students working toward content and process goals. Several issues that can undermine classroom civility are lateness, extended off-task conversations, and long transitions between student-centered and instructor-centered activities. Effectively managing a large POGIL classroom requires cultivating a climate of respect at the beginning and throughout the semester; suggestions for establishing classroom civility and for dealing with specific civility issues follow.

- Begin the first class by facilitating a discussion in which students generate lists of expectations, with rationales, for the course, their instructor, and their peers.
- Establish and enforce ground rules based on expectations.
- Pose a clicker question at the beginning and end of class. Students are awarded participation points and only receive credit if both questions are answered.

- Designate an area in the classroom for late students. Late students go directly to this area, form an ad hoc group, and work on classroom activities without disrupting their established groups who have already begun working on a task.
- Write and use activities that are challenging but not overwhelming in order to keep students on task.
- Recruit enough facilitators so that a group may get help if they become frustrated.
- Choose a technique to refocus students on the instructor at the end of a period of group work by counting down the time remaining (posting a countdown clock) or using a physical signal (hand-raising, a variation of the wave, turning lights off and on, or posting a special slide asking for attention).

Student Buy-In

Students in large classes may be especially skeptical of any non-traditional teaching method. Students may have even selected a large class in order to remain anonymous. Several strategies, listed below, help students understand why the course includes POGIL activities. It may be necessary to reinforce the rationale behind the POGIL method throughout the semester; however, it is probably unwise to represent the POGIL implementation as "experimental." This may suggest to students that their instructor does not know that POGIL is valuable. As such, students may respond in an unpredictable and counterproductive manner.

- Provide examples to demonstrate the need for process skills in the work force, including documentation from industry and professional school representatives regarding the importance of working in teams, developing leadership and communication skills, and learning to think critically.
- Discuss the importance of connecting with other students and forming study groups. In larger institutions students may not know anyone in a large class and may feel intimidated or overwhelmed. Most two-year colleges are commuter colleges, and students "evaporate" when they leave the classroom, also making it difficult for students to make connections with fellow students.
- Maintain a strong, positive attitude, and avoid public, in-class debates with students over the value of POGIL. Invite students to visit office hours if further discussion is desired.

Some students will resist collaboration and guided inquiry at the beginning, but buy into POGIL after a few weeks. Some may only realize the value of POGIL after the course ends, and some may never appreciate it.

Introducing POGIL to Colleagues and Administrators

Students are not the only people in an educational setting who should understand the POGIL method and its rationale. Colleagues, administrators, and parents can be skeptical of collaborative learning in large classes. Like students, not every colleague and administrator will endorse POGIL in large classes. However, the following strategies can be useful in creating buy-in from groups outside the classroom:

- Ask colleagues to come to your classroom.
- Describe the pedagogy, objectives, rationale, and effectiveness of POGIL.
- Share student and faculty anecdotes describing a POGIL classroom experience from students and faculty.
- Present documentation from industry and professional school representatives regarding the importance of working in teams.
- Use documents from the POGIL web site for additional support.
- Give a presentation or workshop to other faculty through the local Teaching and Learning Center on campus.
- Form a core group of like-minded colleagues (inside or outside the chemistry department) to share ideas and frustrations and to provide mutual support.

Curriculum Constraints

Many instructors perceive that content coverage is sacrificed in large classes using POGIL. They worry that, if some content is missed, students may fare poorly in subsequent courses. Others may be concerned that students in a POGIL classroom will get lower scores on common exams. Instead, as shown in Chapters 19 and 20, and in other studies (*19 – 23*), evidence actually shows that students in POGIL classrooms perform as well as or better than their peers from traditional lectures on exams.

A few simple stratcgics hclp students stay on task, thus ensuring that the necessary concepts in a given course are taught (and learned).

- Follow a set schedule. When working on a POGIL activity, announce the actual time when the instructor will solicit responses from individual groups.

- Assign unfinished class work as homework. The instructor can ensure that the work is completed by giving a clicker quiz at the start of the next class period.
- Ask clicker questions at several points during an activity (see Classroom Organization section above).

Recruiting and Training Undergraduate Facilitators

More facilitators are needed in large POGIL classes: one facilitator is needed for approximately every 20 students. Undergraduate students can be recruited to be facilitators, especially at the larger institutions where more facilitators are needed. Students can be rewarded for being facilitators with money (perhaps work-study funds), independent study credit, or regular course credit (i.e., create a "Facilitation" or "Guided Inquiry" course). Undergraduate students are often willing to be facilitators without being paid, as facilitating instruction provides leadership opportunities and/or more in-depth exposure to course content. The following are suggested undergraduate facilitators:

- Students who did well in your course in previous semesters.
- Pre-service teachers who have taken chemistry courses.
- Pre-health science students (pre-med, pre-dent, pre-vet, pre-pharm) who need to review for entrance exams and who need faculty letters of recommendation.
- For two-year colleges, students from neighboring four-year schools.
- Majors looking for service opportunities.
- Students who are active in the student affiliates of the American Chemical Society.
- Entering first-year students who place out of general chemistry based on their Advanced Placement exam scores.

Recruiting undergraduate facilitators may be the easiest part of implementing POGIL in large classrooms. For example, instructors at Portland Community College recruited student facilitators from neighboring colleges selected by their chemistry departments, which granted them upper-division science credits. Subsequently, over 50% of students recruited for the following year volunteered to be peer facilitators with no compensation or credit.

Regardless of the characteristics of the student facilitators, training is necessary. At institutions with established student facilitator training programs, facilitators are trained before the course begins and in weekly meetings throughout the semester. The introduction to POGIL is modeled after the one-day POGIL workshop. Instructors also discuss ethics, dating, gender and diversity, how to dress, how to facilitate learning (coaching vs. explaining), how to facilitate the group process, and instructor expectations. Undergraduate facilitators keep a journal that documents their challenges and successes and then discuss these in the weekly meetings.

An instructor contemplating the implementation of a new facilitator training program should conduct a thorough review of existing programs from a variety of institutions and reform initiatives. Several good places to begin this review are the University of South Florida (*22*), Stony Brook University (*24*), and Peer-Led Team Learning (*25, 26*).

Diversity

Large classes at many institutions are often diverse with respect to age, gender, race, nationality, and interest. Some students may work (at a variety of jobs), have families, or come from an urban, suburban, or rural background. Students have a broad range of aptitudes for chemistry. This diversity foreshadows the diversity students will encounter after graduation in the workplace. Employers expect that students will know how to work with all types of people. Thus, using POGIL in large and diverse classrooms benefits students, although managing groups in such a diverse classroom can be challenging. If possible, the diversity of student facilitators should match the diversity of students in the class, and the groups should have a mix of students (diverse by gender, ethnicity, etc). It may be necessary for facilitators to work closely with diverse groups to help them manage their differences. For example, older students may not like to work with much younger students (who are perceived to have fewer responsibilities or to be less serious).

Conclusion

Instituting POGIL in large classrooms does not have as long a history as its use in smaller classes. There are quite a few examples across science for creation of student interactions in large classrooms. Some of these examples are clearly intended to support student inquiry, articulation of understanding, and testing of ideas. This is the GI (Guided Inquiry) part of POGIL. As such, instructors who wish to use POGIL in large classes may find it easier to tackle the GI part first. One strategy is to replace lecture segments with short group activities in which students examine a model or data and respond to critical thinking and application questions.

What will be the more significant challenge is sustaining the PO (Process Oriented) part of POGIL – the development of explicit group process structures and assessment. As such, the PO part should be addressed after instructors are comfortable with group work and their activities and wish to focus more closely on group interaction and assessment.

References

1. Hillstock, L.G. *A few common misconceptions about distance education*; Report from Association of Small Computer Users in Education, 2005.
2. Imel, S. *E-Learning. Trends and Issues Alert*; ERIC Clearinghouse on Adult, Career, and Vocational Education, 2003.
3. Wilner, A; Lee, J. *The Promise and the Reality of Distance Education*; National Education Association: Washington, DC, 2003.
4. Hedberg, J.G. *Studies in Continuing Educ.* **2006**, *28*, 171-183.
5. Tomei, L. A. *J. Technol. Teacher Educ.* **2006**, *14*, 531-541.
6. Tabatabael, M.; Schrottner, B.; Reichgelt, H. *International J. E-Learning* **2006**, *5*, 401-414.
7. Powell, R.; Keen, C. *International J. Higher Educ. and Educ. Planning* **2006**, *52*, 283-301.
8. Mazur, E. *Peer Instruction: A User's Manual*; Prentice Hall: Upper Saddle River, NJ, 1997.
9. Heron, J. D. *The Chemistry Classroom: Formulas for Successful Teaching;* American Chemical Society: Washington, DC, 1996.
10. Felder, R. M. *Chem. Engin. Ed.* **1991**, *25*, 132-133.
11. Carbone, E. *New Direc. for Teac. and Learn.* **1991**, *77*, 35-43.
12. Cooper, J. L.; Robinson, P. *New Direc. for Teac. and Learn.* **2000**, *81*, 5-16.
13. Wulff, D. H.; Nyquist, J. D.; Abbott, R. D. In *Teaching Large Classes Well*; Weimer, M. G., Ed.; San Francisco: Jossey-Bass, 1987; pp 17-30.
14. Tobias, S. *They're Not Dumb, They're Different*; Research Corporation: Tucson, AZ, 1990.
15. McKeachie, W. J. *Teaching Tips: Strategies, Research, and Theory for College and University Teachers;* D. C. Heath and Company: Lexington, MA, 1994.
16. Hanson, D. M.; Wolfskill, T. *J. Chem. Educ.* **2000**, *77*, 120-130.
17. Hanson, D. M. In *Faculty Guidebook: A Comprehensive Tool for Improving Faculty Performance*; Beyerlein, S. W., Apple, D. K., Eds.; Pacific Crest: Lisle, IL, 2004; pp 305-308.
18. Johnson, D. W.; Johnson, R. T.; Smith, K. A. *Active Learning: Cooperation in the College Classroom*; Interaction Book Company: Edina, MN, 1991.
19. Farrell, J. J.; Moog, R. S.; Spencer, J. N. *J. Chem. Educ.* **1999**, *76*, 570-574.
20. Hanson, D. M.; Wolfskill, T. *J. Chem. Educ.* **2000**, *77*, 120-130.
21. Hinde, R. J.; Kovac, J. *J. Chem. Educ.* **2001**, *78*, 93-99.
22. Lewis, J. E.; Lewis, S. E. *J. Chem. Educ.* **2005**, *82*, 135-139.
23. McKnight, G.; Salem College, unpublished, 2004.

24. Hanson, D. M.; *Instructor's Guide to Process-Oriented Guided-Inquiry Learning*; Pacific Crest: Lisle, IL, 2006; p 31.
25. Roth, V.; Goldstein, E.; Marcus, G.; *Peer-Led Team Learning: A Handbook for Team Leaders*; Prentice Hall: Upper Saddle River, NJ, 2001.
26. Gosser, D. K.; Cracolice, M. S.; Kampmeier, J. A.; Roth, V.; Strozak, V. S.; Varma-Nelson, P; *Peer-Led Team Learing: A Guidebook*; Prentice Hall: Upper Saddle River, NJ, 2001.

Chapter 7

Facilitation: The Role of the Instructor

Vicky Minderhout and Jennifer Loertscher

Department of Chemistry, Seattle University, 901 12th Avenue, Seattle, WA 98122

Facilitation in a POGIL-style course involves instructors guiding students to develop process skills and conceptual understanding while they work in self-managed teams on specially designed activities. In a POGIL classroom, written activities provide content focus, whereas instructor facilitation creates opportunities to improve students' abilities to learn by developing the process skills necessary for learning. Effective facilitation is planned for and improved upon with practice and critical reflection. Preparing a facilitation plan for each class session helps instructors identify the necessary process skills that students must have to learn content and function within a group, anticipate what evidence they would observe if students were utilizing those skills, and make plans for effective intervention to improve those skills. By creating a plan, instructors also have a record of what was planned for and what actually happened in class and can reconcile the two to provide direction for future efforts. The plan also serves to document one's growth as an instructor in active learning approaches and guides the selection of facilitation tools and group interventions to improve student learning.

The trend in education toward learner-centered approaches means that student learning, rather than instructor delivery of education, becomes the central outcome. Empowering learners is challenging and complex work. To deliver on the promise of learning, faculty need to conduct assessment *for* learning rather than just assessment *of* learning. Assessment for learning implies that the assessment is used in a formative manner to understand what process skills the students need work on to master content and guide development of that learning. Assessment for learning can be done in real time in the classroom while a faculty member is facilitating active learning. In this way faculty become researchers in their own classrooms, learning what their students know and can do and what they don't know and cannot do. As a result, faculty can modify instruction in real time or design future instruction to help their students improve their abilities to learn and to apply the knowledge they have learned. Thus the process of facilitation is an important tool for faculty in recognizing and developing the capabilities of their students.

Overview of Facilitation

The term facilitation has a variety of meanings depending on the context. In a POGIL classroom, students work in self-managed teams on specially designed activities that help them master and apply discipline-specific concepts (*1*). While students actively engage in cognitive development associated with the discipline, they also develop essential process skills, including teamwork and communication. In terms of affective development, students learn to manage their frustration with learning and practice being patient with others. Therefore teachers who facilitate learning in environments like POGIL have a unique opportunity to help students develop skills beyond those needed for the practice of the discipline. However, facilitating an active learning classroom is not like lecturing and instructors need to develop different skills to maximize effectiveness in the classroom (*2-5*). Clearly instructors interested in pursuing active learning strategies in their classrooms would benefit from observing others employing these techniques, but that is often not feasible. A goal of this chapter is provide instructors with a framework by which they can plan their classroom facilitation, learn from it, and use it to document their growth as a teacher.

How Facilitation Fits into Curriculum Design

Facilitation and curriculum design are interdependent. Cooperative learning involves many content-free procedures (*6*) associated with creating and implementing the learner-centered environment. Cooperative procedures and practices are part of course design and provide students with opportunities to improve social and affective skills. Along with the cooperative learning procedures, a course designed to increase the level of challenge over time from foundational understanding to higher order skills, such as application and problem solving, creates opportunities for students to improve and build cognitive process skills due to the increasing levels of difficulty and complexity. In addition there is often discipline specific procedural knowledge that must be acquired prior to application of concepts to complex problems. Teachers need to assure that the prior knowledge of lower order cognitive, social and affective skills have been practiced and coached earlier in the term in order for students to experience some success in higher order processes. Still, students often are not able to recognize barriers to their progress as they move beyond simple knowledge and understanding. Therefore, a substantive part of the facilitator role is to help teams identify those skills limiting their progress. This seems a daunting task at first, but over time and with practice, planning, and reflection, a facilitator will know and recognize those skills affecting student progress. Learning how to help students surmount these learning barriers is exciting and worthwhile work.

Profile of a Quality Facilitator

As instructors plan classroom facilitation and act to improve their facilitation skills, some benchmarks are useful. Peter Smith (*7*) has organized the behaviors of a quality facilitator into a table entitled Profile of a Quality Facilitator (see Table I). This table addresses six categories including preparing, assessing audience needs, activity setup, facilitating experience, activity closure and follow-up. Within these categories Smith offers five discrete descriptions of behaviors that serve as valuable guides for improving facilitation skills. It is possible to coach oneself and improve facilitation skills using the descriptors as targets coupled with formative assessment of one's teaching practice. A good facilitator constantly modifies his or her approach based on assessment of facilitation experiences. The facilitation plan discussed in this chapter provides further guidance for teachers to improve in facilitation.

Facilitator Skills

Several specific skills help an instructor exhibit the behaviors of a quality facilitator. A complete listing of and description of facilitation skills is available (*8*), but the most important skills, having the ability to listen and rephrase, asking

Table I. Profile of a Quality Facilitator

Profile of a Quality Facilitator

Preparing	• Develops resources for multiple scripts/tasks. • Designs strong structures through a facilitation plan, a road map. • Predicts the major issues that must be addressed including what "done" looks like. • Prepares background conceptual knowledge. • Defines metrics for project success, such as cost, schedule, performance, or quality.
Assessing audience needs	• Affirms what each brings to the table. • Discovers major issues people are confronting. • Seeks out the outcomes for each person. • Identifies collective outcomes. • Clearly predicts and verifies everyone's role in moving along the road map.
Setup	• Clarifies expectations. • Creates a framework for the process; describes the road map and major milestones. • Establishes teams. • Motivates individuals for the experience. • Performs risk assessment and predefines risk management.
Facilitating experience	• Constantly transfers ownership to participants. • Actively assesses progress of individuals and teams. • Constructively intervenes on process issues, not content. • Continuously raises the bar to challenge participants. • Monitors objective metrics and actively acts on data to ensure success.
Closure	• Stops activity at the top of the production curve. • Requests each team representative to summarize issues, good and bad. • Does a perception check for consensus within each team. • Makes sure that each issue has an owner and due date to ensure resolution. • Insists on assessment of learning processes.
Follow-up	• Makes sure team members achieve individual/collective outcomes. • Accepts constructive criticism and promises action toward improvement. • Ensures that all data is collected for participant reflection. • Reinforces negative and positive issues as equally important. • Clarifies the next step in the process.

SOURCE: Reproduced with permission from reference 7. Copyright 2005 Pacific Crest.

critical questions, and recognizing emotions, will be described here in some depth.

Two important skills are the abilities to listen and rephrase. These skills are essential for the facilitator in order to discover and clarify major issues students are confronting, to assess progress of individuals and teams, to check for consensus within the team, and to effectively intervene. Listening and rephrasing are also important skills for students in an active learning environment. Therefore, when used by the instructor during class, these skills not only improve the quality of facilitation, but also have the added benefit of modeling useful process skills for students. Rephrasing is a particularly relevant skill to model since it is a way for shy or less confident students to participate without presenting or defending their own ideas. There are no wrong answers when one is rephrasing (*9*). By using rephrasing as a way to interact with groups, an instructor can gather essential information needed for effective facilitation and at the same time model this important skill for the students.

Once the facilitator gets a sense of student learning issues through listening and rephrasing, the facilitator can model another essential skill, the ability to ask critical questions. The active nature of the POGIL classroom gives instructors a window into their students' thought processes and therefore puts instructors in a position to consider if an intervention would help promote student learning. Instead of intervening directly on content or process, highly competent facilitators use questions to help students identify and clarify the relevant issues affecting learning. The ability to ask pertinent questions is also important for students working within teams, giving the facilitator another opportunity to model an essential skill for students. Furthermore, the frequency, relevance and cogency of student questions are indicators of critical thinking (*10*). Therefore, by paying attention to student questions, instructors can ascertain the level of student understanding.

Recognizing emotions is a facilitator skill that is extremely useful in learner-centered classrooms. Learner-centered approaches may be particularly difficult for some students. According to Weimer, "Our students are passive, disconnected, not always responsible, dependent and not very confident learners. ....some strategies require a level of intellectual maturity that they may not possess at the beginning." (*11*) Learning is in fact hard work, especially since we are now expecting more of the students on a regular basis. Facilitators need to monitor affective behaviors since significant negative emotions will impede learning (*12-14*). Facilitators can monitor the emotions of groups and individuals in the classroom by quickly skimming the classroom environment and being sensitive to body language (*15*). Davis provides a useful list of the signs of conflict and apathy in groups (*16*) and Johnson and Johnson provide a wealth of information in developing group skills (*17*). If a problem is identified, the facilitator can intervene before a negative situation goes too far.

Creating a Facilitation Plan

A facilitation plan serves three purposes: to guide a successful faculty performance before, during and after active learning; to form a permanent record of what was attempted; and to improve future efforts. Using a consistent, written format is a good way to ensure that these purposes are met. Published facilitation plan templates are available (*18*), but faculty may choose to develop a form that addresses specific priorities.

To illustrate the elements and implementation of a facilitation plan, an example class session of a senior level biochemistry course will be used. The class day was three weeks into the term, one week before the first exam and the students had worked in the same groups since the beginning of the course. Students were told that the session would be the first of two sequential problem solving days in which they will apply their newly acquired knowledge about proteins and biochemical techniques (see Figure 1). The subsequent session would be devoted to problem solving in which teams earned points for their problem solution. Problem solving in the POGIL classroom refers to relatively complex questions requiring higher level thinking skills for which the solution is not immediately apparent to the student. Students are therefore required to discern the exact goal of the task and use judgement and their new knowledge to propose a reasonable approach. During this activity, groups were asked to propose experiments. If the experiments utilized the appropriate technique and included proper controls, teams were given experimental results by the instructor. Poorly designed experiments yielded the statement, "the data are uninterpretable". The experimental results obtained were used by the team to create the final response for the problem.

Before Class

The elements of the facilitation plan that are completed by the facilitator before the class session are: activity outcomes, type of activity, team roles, pre-assessment, activity set-up, skills needed to perform the activity, evidence of skill usage, anticipated interventions and closure of the session. In preparing a plan for an activity one has taught before, reviewing the notes from last year's facilitation plan is a good starting point since similar issues are likely to arise again. Since the notes include whether or not the learning outcomes were met in the prior offering, one will know if modifications are needed. For the example class period, the notes revealed that students got very frustrated during this day, but that over half the teams followed the provided problem solving method and were able to interpret their results and get a memo written.

MEMO FROM THE BIG BOSS
The person who can get the most information about the structural nature of LWBGase <u>in the next 24 hours</u> will get stock when the company goes public next week. Send all your information and data to the head honcho of the structural group for evaluation.

You have at your disposal some standard biochemicals, an inventory of standard proteins, and various chromatography and electrophoresis equipment. Explain what you would do, what data you would obtain, and, once you have analyzed the data, what the data suggests about the structural nature of this protein.

Figure 1. Problem Solving Activity Example

Activity Outcomes

In selecting activity outcomes as part of a facilitation plan, it is useful to consider social and affective outcomes along with the cognitive outcomes, with which most faculty are comfortable. Keep in mind that one of the reasons to use cooperative learning rather than lecture is to develop skills in these other domains. There is much research that describes the links between intellectual, social and emotional elements of learning (*19,20*). Therefore, being explicit about social and affective outcomes in addition to cognitive outcomes is important for both facilitators and the students.

The outcomes for the example activity included a cognitive, a social and two affective goals. The cognitive outcome was for the students to apply knowledge acquired during the first three weeks of class by using a generic problem-solving method. The social outcome was to develop teamwork skills needed to solve a complex problem. Since this activity was a precursor to problem solving for team points, students were able to practice teamwork skills in a low-risk environment. There were two affective outcomes. First was to develop students' ability to manage frustration since the problem was very undirected and their knowledge was new. The second affective outcome was building student buy-in to both the class structure and the problem solving method.

Activity Type

Identifying activity type can help the facilitator recognize special challenges that may arise during a given type of activity. The activity type for the example class session was not guided inquiry, class discussion or laboratory, but problem solving and represents a situation that the students might encounter after graduation.

Activity Roles

Facilitators should plan team roles before the activity to enhance the functioning of the groups (*21*). The roles selected for the example activity were manager, recorder, reflector and optimist. In this type of problem solving activity, the manager needs to have a strong presence due to the anticipated team frustration. The role of the reflector is to monitor team process and help the team make adjustments to improve their team performance. Because of the importance of the reflector in the example case, this role was narrowed to focus on monitoring the stress level of the team and to track the usage of the problem solving method. Narrowing the reflector's role is a technique facilitators can use to help monitor specific issues within the team (*8*). The recorder's role was used to keep track of the team ideas, manage the data obtained from the "experiments" and ultimately create the memo. The fourth role used in the example activity was that of the optimist, who offers continuous encouragement. The optimist role was useful considering the stress the students experienced during this activity. Of course the selection of roles will depend on the activity. Other roles are designed as needed to support the learning outcomes.

Pre-Assessment

In planning for pre-assessment, facilitators determine whether there is prior knowledge that the participants need to know in order to construct the new knowledge addressed in the activity. Some faculty might choose to administer a quiz, whereas others might call attention to the prior knowledge by having a pre-activity class discussion. The pre-assessment for the example activity was the completion of an assigned problem using the problem-solving method. This assignment assured exposure to and practice with the method prior to class.

Activity Set-up

The script for the set-up of the activity is designed in advance. Being explicit in your communication with the students is critical (*22*). For the problem-solving example, the script included a description of the learning outcomes, mention of the real life nature of the problem situation, the importance of following the problem-solving method, and the activity-specific duties for the usual roles. The new role of the optimist was discussed since it had not been used in class up to that point. In addition, the groups were told that for this activity they will "run" several experiments and if each experiment is executed correctly they will get data from the instructor in order to write their memo. The recorder was instructed to keep track of the data. Finally the class was instructed that proposed experiments not "run" correctly will yield results that are not interpretable.

Learning Skills Needed

One of the most important aspects of planning before one facilitates an activity is to identify the skills needed to successfully achieve the learning outcomes of that activity and to anticipate evidence of their usage (*18,23*). It is usually best to focus on the most significant skills required and work to strengthen those rather than trying to strengthen all skills needed for the activity. By considering the evidence of skill usage before class, the facilitator can quickly monitor a teams' progress by watching for the evidence outlined in the facilitation plan. For the example activity, the first outcome was using a defined method to guide student problem solving. Evidence of using the problem solving method would be written statements in the recorder's report or verbal discussion of method steps. In addition, the facilitator could check in with the reflector, whose job was to monitor use of the method.

Another important skill needed for the problem solving day was managing frustration. Planning strategies to identify and mitigate frustration is crucial in good facilitation, since negative emotions impede learning (*20,24*). The facilitator must be prepared to make a judgment about whether or not the level of frustration is actually impacting learning and be ready to act with a predetermined mode of intervention (*25*). Hanson suggests several generic questions, (*1*), that can serve numerous stressful situations: "What are you doing? Why are you doing it? How will it help?" Johnson and Johnson provide suggestions for teaching social skills (*26*). Giving teams a choice is also effective in relieving frustration (*27*). For example, asking the team if they would like some consulting gives the team an opportunity to choose to ask for help and keeps the balance of power in their hands.

Closure

An important part of effective facilitation is to capture and revisit the learning gains and insights made in content and process areas so that students can plan improvements (*28*) and develop a genuine sense of accomplishment. Therefore, the plan for closure of the class period should help solidify the learning gains made during class and reveal how the outcomes have been achieved. Because students are often less aware of gains made in process areas, a plan for closure should pay particular attention to these areas (*24*). Closure is oftentimes difficult since students want to continue discussion; therefore, plans for closure need to include aspects that the students will find more useful than their own discussions (*11*). For the problem solving example closure was extremely difficult since teams that have not completed a memo do not want to quit. The plan therefore was to present the students with the data correlating effective use of the problem solving method with completion of the activity and to ask them to reflect on why those teams who used the method efficiently were able to finish.

During Class

A facilitation plan is created before class and executed during class. As class proceeds the facilitator has the opportunity to collect data about how much time was used for each component of the activity, (pre-assessment, set-up, activity, closure) and about the effectiveness of each component. As students complete the activity, the facilitator monitors the progress of each team and collects real-time data on the interactions occurring within the team and evidence of the skill usage identified prior to class. Content outcomes, the responses to activity questions, are relatively easy to observe. Taking notes about the student difficulties and successes observed during class will provide an invaluable resource to improve facilitator skills even if little effective planning was done prior to the class session. For the example of the problem solving class, the data revealed that many teams do not follow the problem solving method provided. As a result, these teams experienced a disorganized group process and frustration just around the corner.

Interventions

Intervening in a way that empowers students and promotes learning can be one of the most challenging aspects of facilitation. The more one anticipates problems and plans solutions, the more effective the intervention will be. For an example of effective intervention, recall that evidence for student performance in the example activity was effective use of a problem solving method. One key to effective intervention in this case was to realize that the affective outcomes of managing frustration and promoting buy-in are closely linked to effective use of the problem solving method. If a group was becoming too frustrated and was not managing it well, a simple intervention, like asking the team if they would like some consulting, might redirect their energies productively. If they answered yes, a simple observation about their process, e.g., "I do not see a defined problem statement" could refocus their conversation. Experience has revealed that allowing teams to struggle is useful for eventually getting them to use the problem solving method. Since using the method is the outcome, rather than actually solving the problem, most groups were left to struggle as long as their level of frustration was appropriately being managed.

Once the team is calmly talking the intervention is over. The facilitator should revisit the team in a few minutes and collect data on the effects of the intervention on the group. These observations are essential to help build strong facilitation skills (*25*). Finally, to close the loop, it is best if the students reflect on their learning issues as soon as possible so that they can take action to remedy it in the future (*29*). The instructor can help prompt this thinking by asking students to state what they will do in the future. In this way the instructor helps the students build the self-monitoring skills so important in expert thinkers and problem solvers (*30*). As stated by the authors of *Assessment for Learning* (*31*), "feedback can only serve learning if it involves both *evoking* of evidence and a response to that evidence by *using* it in some way to improve learning." This thinking about thinking, meta-cognition, prompted by the instructor can help students understand how to learn better (*23*) and can directly help with buy-in to the course structure.

Interventions on content (providing answers) rather than process can have undesirable effects (*32*). The facilitator risks reestablishing herself as the content expert and devalues the expertise and abilities of the group. This can diminish student self-confidence and increase dependence, as students will look to the facilitator rather than themselves for answers in the future (*33*). Furthermore, giving groups answers often prevents the facilitator from discovering the fundamental process difficulties the group is experiencing.

Closure

Closure of the activity is one of the hardest aspects of facilitating active learning. Capturing learning discoveries or insights is important for future learning (*28*), but insights are undervalued by most students especially those new to active learning experiences (*13*). Naturally, student focus on content because that is what will be on the test. However, improving student learning and process skills will have lasting effects. Therefore each teacher needs to strike a balance that feels satisfying to students, while continuing to raise awareness of the valuable skills being practiced and the insights that were made during class.

Closure planned for the example class was to present the students with the data correlating effective use of the problem solving method with completion of the activity and to ask them to reflect on why those teams who used the method efficiently were able to finish. One team observed that the method helped them stay organized and focus their conversation rather than having four different approaches. Students also recognized that the undirected problem represents a realistic situation that they may encounter in the workplace. Through reflection students realized that most difficult part of the task given by "the boss" was not finding an answer, but determining the experimental questions. These ideas are used to reinforce for students why the course is taught this way, why they are asked little recall on tests, and why they need to be able to solve undirected, complex problems.

After Class

The notes collected by the facilitator during class, as well as recorder's and reflector's reports, represent a wealth of information from each class period. These data can be used to reconcile what happened in the class with what was planned. An assessment report like that described by Wasserman and Beyerlein (*34*) can organize the process of reflection on practice. If the outcomes were met, it is useful to identify factors contributing to this success so that the facilitator can incorporate successful strategies into future class sessions. If the student learning outcomes were not met, an appraisal of activity components (outcomes, prior knowledge, roles, activity type, closure) is necessary to plan for improvements. In determining whether the outcomes were met, remember that content outcomes are part of the designed activity while process outcomes depend on structural elements of the course including activity type and a facilitation plan. Without repeated reflection and planning, successes might be forgotten and the same mistakes might be made each year. Therefore, making a plan to improve the next class period and future facilitation efforts closes the loop in facilitation and provides continuing evidence of growth as an instructor in active learning strategies.

Summary

While Johnson, Johnson and Smith coined the term "guide on the side" (35), we have tried to articulate what it means to be the guide on the side in the POGIL active learning classroom. Not surprisingly, fulfilling this role does not come without careful planning, execution, and reflection on the part of the instructor. Since the goals of most POGIL practitioners is to help their students develop the process skills and conceptual understanding essential for life-long learning, we believe that the time invested to develop facilitation plans and continually improve one's performance as an instructors well worth the effort. By removing the communication barriers present between students and instructors in traditional classrooms, POGIL puts the instructor in a position to help students improve their learning processes and build transferable skills in real time. By viewing themselves as active facilitators who plan for and anticipate learning challenges and opportunities prior to class, instructors can take full advantage of the interpersonal dynamics in a POGIL classroom to positively shape student learning.

References

1. Hanson, D. *Instructor's Guide to Process-Oriented Guided-Inquiry Learning;* Pacific Crest: Lisle, IL, 2005; p 46.
2. Johnson, D. W.; Johnson, R. T.; Smith, K. A. *Cooperative Learning: Increasing College Faculty Instructional Productivity*; The George Washington University, Graduate School of Education and Human Development: Washington, DC, 1991.
3. Mazano, R. J.; Pickering, D. J.; Pollock, J. E. *Classroom Instruction that Works: Research-Based Strategies for Increasing Student Achievement*; Association for Supervision and Curriculum Development: Alexandria, VA, 2001.
4. Millis, B. J.;Cottell Jr., P. G. *Cooperative learning for higher education faculty*; Oryx Press: Phoenix, AZ, 1998.
5. Weimer, M. *Learner-Centered Teaching*, Jossey-Bass: San Francisco, CA, 2002.
6. Millis, B. J.; Cottell Jr., P. G. *Cooperative learning for higher education faculty;* Oryx Press: Phoenix, AZ, 1998; p 69.
7. Smith, P. In *Faculty Guidebook: A Comprehensive Tool for Improving Faculty Performance, 2nd Ed.;* Beyerlein, S. W.; Apple, D. K.; Ed; Pacific Crest: Lisle, IL, 2005; pp 137-140.
8. Minderhout, V.; Smith, P. In *Faculty Guidebook: A Comprehensive Tool for Improving Faculty Performance, 2nd Ed.;* Beyerlein, S. W.; Apple, D. K.; Ed; Pacific Crest: Lisle, IL, 2005; pp 157-160.
9. McKeachie, W. J.; Svinicki, M. *McKeachie's Teaching Tips: Strategies, Research and Theory for College and University Teachers, 12th Ed.;* Houghton Mifflin: Boston, MA, 2006; p 47.
10. Postman, N.; Weingartner, C. *Teaching as a Subversive Activity;* Delacorte Press: New York, NY, 1969; pp 36-37.
11. Weimer, M. *Learner-Centered Teaching;* Jossey-Bass: San Francisco, CA, 2002; p 153.
12. Love, P. G.; Love, A. G. *Enhancing Student Learning: Intellectual, Social and Emotional Integration*; The George Washington University Graduate School of Education and Human Development: Washington, DC, 1995.

13. Weimer, M. *Learner-Centered Teaching;* Jossey-Bass: San Francisco, CA, 2002; pp 149-61.
14. Millis, B. J.; Cottell Jr., P. G. *Cooperative learning for higher education faculty;* Oryx Press: Phoenix, AZ, 1998; p 39.
15. Apple, D. K.; Duncan-Hewitt, W.; Krumsieg, K.; Mount, D. *Handbook on Cooperative Learning*; Pacific Crest: Lisle, IL, 2000.
16. Davis, J. R. *Better Teaching, More Learning;* American Council on Higher Education and Oryx Press: Phoenix, AZ, 1993; pp 260-61.
17. Johnson, D. W.; Johnson, F. P. *Joining Together: Group Theory and Group Skills, 8th Ed.*; Allyn and Bacon: Boston, MA, 2003.
18. Minderhout, V. In *Faculty Guidebook: A Comprehensive Tool for Improving Faculty Performance;* Beyerlein, S. W.; Apple, D. K.; Ed.; Pacific Crest: Lisle, IL, 2005; pp 145-148.
19. Johnson, D. W.; Johnson, R. T.; Smith, K. A. *Cooperative Learning: Increasing College Faculty Instructional Productivity;* The George Washington University, Graduate School of Education and Human Development: Washington, DC, 1991; pp 27-55.
20. Love, P. G.; Love, A. G. *Enhancing Student Learning: Intellectual, Social and Emotional Integration;* The George Washington University Graduate School of Education and Human Development: Washington, DC, 1995; pp 29-41.
21. Smith, P. In *Faculty Guidebook: A Comprehensive Tool for Improving Faculty Performance, 2nd Ed.*; Beyerlein, S. W.; Apple, D. K.; Ed.; Pacific Crest: Lisle, IL, 2005; pp 207-210.
22. Weimer, M.; *Learner-Centered Teaching;* Jossey-Bass: San Francisco, CA, 2002; p 157.
23. McKeachie, W. J. In *New Directions for Teaching and Learning: Learning, Cognition and College Teaching No. 2;* McKeachie, W. J.; Ed.; Jossey-Bass: San Francisco, CA, 1980; pp 86-88.
24. McKeachie, W. J.; Svinicki, M.; *McKeachie's Teaching Tips: Strategies, Research and Theory for College and University Teachers, 12th Ed.;* Houghton Mifflin: Boston, MA, 2006; p 54.
25. Leise, C.; Smith, P. In: *Faculty Guidebook: A Comprehensive Tool for Improving Faculty Performance, 2nd Ed.;* Beyerlein, S. W.; Apple, D. K.; Ed.; Pacific Crest: Lisle, IL, 2005; pp 149-152.
26. Johnson, D. W.; Johnson, F. P. *Joining Together: Group Theory and Group Skills, 8th Ed.;* Allyn and Bacon: Boston, MA, 2003; pp 53-55.
27. Costa, A. L.; Garmster, R. J. *Cognitive Coaching;* Christopher-Gordon Books: Norwood, MA, 1994; pp 143-150.
28. Johnson, D. W.; Johnson, R. T.; Smith, K. A. *Cooperative Learning: Increasing College Faculty Instructional Productivity*; The George Washington University, Graduate School of Education and Human Development: Washington, DC, 1991; p 69.

29. Black, P.; Harrison, C.; Lee, C. Marshall, B.; Wiliam, D. *Assessment for Learning*; Open University Press: Berkshire UK, 2003.
30. Kurfiss, J. G. *Critical Thinking: Theory, Research, Practice, and Possibilities*; Association for the Study of Higher Education: Washington, DC, 1988.
31. Black, P.; Harrison, C.; Lee, C.; Marshall, B.; Wiliam, D. *Assessment for Learning*; Open University Press: Berkshire UK, 2003; p 122.
32. Hanson, D.; Wolfskill, T. *J. Chem. Ed.* **2000**, *77*, 120-130.
33. Mazano, R. J.; Pickering, D. J.; Pollock, J. E. *Classroom Instruction that Works: Research-Based Strategies for Increasing Student Achievement;* Association for Supervision and Curriculum Development: Alexandria, VA, 2001; pp 96-99.
34. Wasserman, J.; Beyerlein, S. In *Faculty Guidebook: A Comprehensive Tool for Improving Faculty Performance, 2nd Ed.*; Beyerlein, S. W.; Apple, D. K.; Ed.; Pacific Crest: Lisle, IL, 2005; pp 243-244.
35. Johnson, D. W.; Johnson, R. T.; Smith, K. A. *Cooperative Learning: Increasing College Faculty Instructional Productivity;* The George Washington University, Graduate School of Education and Human Development: Washington, DC, 1991; p 81.

Chapter 8

What Do Students Experience during POGIL Instruction?

Patrick L. Daubenmire[1] and Diane M. Bunce[2]

[1]Center for Science and Math Education and Department of Chemistry, Loyola University at Chicago, 6525 North Sheridan Road, Chicago, IL 60626
[2]Department of Chemistry, The Catholic University of America, 620 Michigan Avenue NE, Washington, DC 20064

The adoption of POGIL as an instructional practice takes time, energy, and often, a restructuring of one's teaching philosophy. In order to adopt such a program, it is important to understand how students may be affected. This chapter describes research which observed, analyzed, and interpreted students' behaviors during a POGIL general chemistry course. Students were involved in the analysis. Results show that students experience Phases and Bridges of Learning during POGIL, which have benefits to their learning of chemistry. Even though such benefits are present, not all POGIL classes are identical. Small variations in implementation can create differences in student outcomes.

POGIL and Theories of Learning

Process Oriented Guided Inquiry Learning (POGIL) is more than a set of activities for student to work through; it is also an instructional practice which integrates activities into a research and theory-based instructional framework to assist students' learning. According to constructivism, meaningful learning takes place when students are actively involved and discussing ideas with other students. (*1-3*). The learning cycle describes how students process information into knowledge and understanding (*4*). POGIL is based on both aspects of learning – working with others to build knowledge through a process of exploration, concept invention, and application of information (*5, 6*).

What Does POGIL Deliver to Students?

In general, POGIL activities (often referred to as ChemActivities) are based on the learning cycle (4,*5*). ChemActivities provide students with data and/or pertinent initial information. Through questions that emphasize critical thinking and application, these activities guide students in their development of chemical concepts through authentic analysis and use of the provided data and information. The instructional design of POGIL has students, each with an assigned role, working in groups with clear guidelines to accomplish the day's objectives. The instructor monitors students' progress and intercedes when necessary. This intervention is commonly used to redirect a group that has lost focus or to emphasize an important point or insight within the activity.

Initial Indicators of Students' Achievement when Instructed with POGIL

POGIL instructors have reported decreases in withdrawal, deficiency, and failure rates (WDF) when using POGIL compared to their own use of traditional lecture approaches (*6*). This WDF decrease is consistent with comparative studies of cooperative versus traditional lecture approaches (*3*, *7*). There is still a need to explain and understand the utility of POGIL and to describe exactly what students experience during POGIL, which includes recognizable student behaviors during a POGIL class and how these behaviors align with theories of constructivism, social interaction, and the learning cycle.

What Do Students Experience during POGIL Instruction?

This chapter provides a synopsis of research which investigated student behaviors during POGIL instruction (*8*). A critical component of the study was the inclusion of students' perspectives of their learning during POGIL through surveys and interview sessions. Research was conducted in two second-semester POGIL general chemistry classrooms. Forty-four student participants were characterized at the beginning of the semester according to demographics (survey), confidence in chemistry ability (survey), and logical reasoning ability (Group Assessment of Logical Thinking (GALT; *9*)). In the demographic survey students provided information on their math and science background and their previous experience in POGIL courses. The Confidence in Chemistry Ability survey asked students to rate their ability to succeed in chemistry in four areas: general confidence to succeed, confidence to handle difficult chemistry *problems*, confidence to handle difficult chemistry *concepts*, and their confidence for success in previous math and science courses. No significant differences of characterizing principles, such as amount of previous POGIL experience and logical reasoning ability, were found between the two participating sections.

Categorizing Student Behaviors

In order to observe a reliable sample of student behaviors in each section, two student groups, randomly assigned by the participating instructors, were videotaped once each week during the 14-week semester. Using a developed coding scheme, researchers observed and analyzed student behaviors during instruction. Initial categories of student behaviors were noted and described in the following way:

1. Students were observed checking responses to questions with one other student in the group. Characteristic phrases which acted as indicators of this category included, "What did you get for answer #2?", "We have different answers for #3, how did you get your answer?"
2. Students discussed their ideas about the information or underlying chemical concept(s). Characteristic phrases which acted as indicators of this category included, "I am not sure I agree. Here is what I think the data says," or "I think we are missing something. This answer does not look right."
3. Students went back over the ideas of their discussion or written answer. Characteristic phrases which acted as indicators of this category included: "How did we get this answer?" or "What did we do when we solved problem #3?"
4. Some interactions were observed in which a student offered to help another student, who had not directly asked for help. A characteristic phrase which acted as an indicator of this category was: "You seemed confused on this problem, can I help?"
5. The group interacted with the Professor. The need for this interaction was determined either by the Professor or the group members.
6. There was significant interaction in which one member of the group who was confused directly asked for help from another group member.

During week five, participants in groups videotaped during week four were invited to participate in interviews about their experience. During the interview session, students viewed selected segments of their own group and participation in class. Students interviewed were asked to describe what was happening in terms of their learning during the segments. The following are responses from students as they relate to each category given above:

1. Checking answers

   Student 134:
   If it is an easier activity that most people kind of understand, we sometimes kind of do it all on our own and then compare answers.

   Student 209:
   When you're doing these questions you have to double check with the people you're with. I could be totally sure of something and then I'd answer it and my group would say 'no, you're completely wrong.' Rather than make a big huge mistake on the test [because of not checking answers], this way, I make a mistake and I learn from it.

2. Discussing as a group

   Student 205:
   If one is by themselves, you don't have anyone there to ask a question, discuss something, have them help you arrive at the right answer, you're just sitting there with your textbook, and your textbook isn't a very moving, live, helpful resource. So basically you can't ask a question. . . It's the dialogue, the conversation that helps people work, I think.

   Student 135:
   Once everybody starts talking, you learn better and quicker. If I would know something and someone else was having trouble, I would explain it and that would help me learn as well as them. And they could do the same for us with someone else in the group.

   Student 134:
   [W]hen the whole group talks about it and stays together, that's a lot easier to figure things out. Because if I have one idea and someone else has another idea, and then somebody else has an idea, we can all say this and somehow put it together.

   Student 219:
   It's just the interaction part, when you hear someone else like throw out an answer that's completely different from yours,

then you're like "what?" You try to just think about it in a different way.

3. Reviewing ideas or answers

   Student 134:
   [S]ometimes we think we have it right, then the Professor comes over and says you need to take a look at number two because that's not quite right. So we go back and reevaluate it.

   Student 202:
   I think the thing I like the most about it is that since you go over it usually a couple of times, it kind of reinforces the things.

4. Ensuring all understand

   Student 212:
   I've learned things and maybe I don't remember them as well, and so, in asking somebody [if he/she needs help] and having to explain it to them, helps bring it back and makes me think about it more.

5. Interacting with the Teacher

   Student 135:
   The times we interact with the Professor are either when he sees something we did wrong, or, if we can't figure out something, we ask him. It happens fairly often.

6. Requesting help from another student

   Student 123:
   When you see somebody [who is successful], you're more prone to pick up what they're doing and it's so much better than watching a teacher. The role of the teacher is so much different than the role of a tutor, a friend trying to help you.

Coding of videotapes was refined in order to integrate students' descriptions of their own behaviors as provided during the interviews.

Characterization of Student Behaviors as Phases and Bridges of Student Learning during POGIL

Analysis of students' descriptions and research observation led to characterizations of student behavior during POGIL. These characterizations have been categorized into four Phases and two Bridges:

Phase I – Compare and Contrast Phase
Phase II – Group Interactive Phase
Phase III – Confirmation Phase
Phase IV – Outreach Phase
Mentor Bridge
Tutor Bridge

Phases

Phase I interactions involved one-on-one interaction to check initial answers each student had written while working individually on initial questions. Inconsistency in these answers created a need to discuss solutions.

During Phase II interactions students discussed their thoughts and ideas about the topic more extensively. After reaching agreement about their answers, students often had to review their thought processes.

In Phase III interactions, students reviewed a previous development of an answer. The need for this type of interaction arose because a Phase II interaction has been involved and they retraced their steps to ensure understanding, or an inconsistency was discovered in subsequent work or was pointed out by the professor. Phase III interactions were concluded when students decided to stay with their current reasoning or change their ideas with new or refined solutions.

Phase IV interactions were entered when an individual group member, apparently confident in his or her own understanding, offered assistance to other group members perceived to be struggling. The member perceived to be struggling had not directly asked for help.

Bridges

Two additional interactions have been characterized as bridges because they were observed to assist student(s) in transitioning from one phase to another. The tutor bridge occurred when one student directly asked another student for clarification or explanation. This interaction was more than a checking of responses as in Phase I. It involved the student who did not understand asking for a one-to-one explanation of an idea. This bridge often helped the struggling student transition from Phase I into Phase II.

The mentor bridge occurred when the group needed guidance because they were not progressing. This occurred either by the request of the group or by the choice of the professor to intercede. This interaction involved questions or statements from the professor which redirected the group. This assisted a group's transition from Phase II into Phase III.

To verify the presence of these Phases and Bridges, additional coders were trained to observe videotapes. Interobserver reliability (85%) supported the presence of Phases and Bridge (*10*).

How Do Students Perceive Their Experiences during POGIL?

Even though students' descriptions were a significant component of the Phases and Bridges characterizations, there was a need to validate the characterizations. A second set of interviews was scheduled. Students once again viewed segments of their group's interactions during a prior class session. Following the segment, the researcher described the Phase and students were asked to comment on the description and type of interaction.

Comments about the Phases

For Phase I, The Compare/Contrast Phase, all 14 students interviewed indicated that Phase I occurred regularly, and it was correctly characterized. Some responses from students on this Phase included:

> *Student 202*:
> I think it does help [to compare and contrast answers] . . . If somebody else has the same thing as you, even if it's the wrong answer, it's like you're getting the same wrong answer. So it's kind of comforting. Like okay – why are we getting this wrong answer or why are we getting the right answer. I think it helps.

> *Student 207*:
> Well, I think if you got it right it just reconfirms what you already know, but if you got it wrong then somebody can explain it to you, why you did it wrong. I think more so than the lecture environment. Somebody can explain it to you. Cause if you're just asking somebody and they're trying to take notes, then they don't have the time to go through it with you, but here at least somebody can go through it with you and explain why it's wrong and tell you how to fix it.

All students interviewed confirmed the presence of Phase II, The Group Discussion Phase. Regarding its presence, some students responded:

> *Student 131*:
> I guess it goes back to more like two heads are better than one, if you have more people working on the problem you have more input. Someone will come at it at a different angle than everyone else and just having someone else to bounce ideas off of. It seems more beneficial to have other people to help you with it. . . . I think with a lecture it's kind of a one on one even though there are other people in the room. It is basically between you and the Professor. But then, with this, you have pretty much the four people in your group, the Professor is always walking around so you can draw him in at some point if you need to get his input.

> *Student 132*:
> If you were just to do it on your own then you would just come to a conclusion. You wouldn't have anybody questioning. You'd be just like, okay, I think the answer is this – and it's like you might be right or there might be a different way to approach it, or there might be a quicker way to approach it. You might not be completely right or you might have the wrong reasoning. If somebody lectured . . . which I [had] last semester. Somebody might say, well this rate equals this rate. Okay, I understand 'cause somebody just did it for me. They just went through the steps. That doesn't mean you can necessarily do it yourself. You might be able to change some numbers and regurgitate it on the test or something, but I don't think you would get the concepts as much as learning like being in a group.

> *Student 207*:
> Usually, someone else thinks of something different. Tom sometimes, he'll bring up something way more complicated than it needs to be, but I think it helps, cause sometimes he's on the right track. We have better chances. Four minds are better than just one. You have better chances of bringing up the right thing.

For Phase III, The Confirmation or "Rehearsal" Phase, 10 of the 14 students interviewed confirmed its presence, two said it sometimes occurs, and two did not clearly confirm or reject its presence.

> *Student 138*:
> [T]o go through and not reflect on what you did minutes earlier would be bad. I think it's (Phase III) an important issue because it ties all the themes together so that when you walk out of there after having done this ChemActivity, I remember them all, they all fit together.

> *Student 131*:
> It happens sometimes, but I wouldn't say it happens all the time. Maybe if like two people in the group come up with one answer and two people come up with another answer then we'll go over it a couple of times with the whole group and maybe bring the Professor in and get his input on it and re-evaluate what we were thinking based on what he said . . . [P]robably the four people would come up with four different answers so then you have to kind of bring those four answers together and so I think that is where it's really beneficial because chances are each of those four answers may have something that isn't quite right with it. So by bringing the four people together you get a better picture as to what the actual answer is. That's pretty much when we move on as if everyone has the same answer and they understand how we got that answer.

For Phase IV, The Outreach Phase, nine students said it occurred, two said it did not occur, and three gave no response that was either a clear confirmation or rejection of the Phase.

> *Student 207*:
> I think maybe if you can explain it to someone in the right way then you know it. That's what teachers have always told me. So I feel like if I talk to somebody and they're listening and I explain it to them, that's giving me confidence to be, like, I can explain this on the test.

> *Student 212*:
> [S]ay I picked everything up really fast. I'd still like to go back and help people because I think that like I'd get more from it than if I'd just closed my book and enjoyed myself and let them work on it. I get more from it if I have to teach them. It imprints it on my brain more. I guess I have a better memory of it if I discuss it.

> *Student 138*:
> I ask them – do you understand how we got that? – or – do you understand where that's coming from – a lot of times they'll be like I get this part, but I don't understand this part of it. Then we try to explain it to each other so basically we teach each other. And [when we] teach somebody else, we're learning it better ourselves as well.

These responses were considered to validate the descriptions and presence of Phases and Bridges.

Additional analysis of time spent in phases and bridges was also conducted. Students spent 46% of instructional time interacting in the Phases and Bridges. Of the remaining time, 25% was used to take daily quizzes or listen to instructions or an overview of the activity, and 29% was used reading information in the activities and forming initial answers or writing down responses following Phase II and/or Phase III interactions.

Do Phases and Bridges Benefit Students' Learning?

Much more research must be conducted to gain an acceptance that students learn better in POGIL because they experience Phases and Bridges of Learning. Four measures, however, at this point support this notion:

1. Previously published results (*6*) demonstrate a decrease in W, D, and F rates for the same professors who had used traditional lecture approaches;
2. Student comments during interviews indicate that Phases and Bridges were present and that they were helpful to their learning;
3. The confidence-in-chemistry survey was administered again at the end of the semester. Post-confidence survey results demonstrated that after having the POGIL course, students viewed groupwork in previous math and science courses to be less effective than when they began the course. To investigate whether these perceptions were lasting, the survey was administered again two-and-a-half years after participating students had POGIL general chemistry. 45% of the original sample responded and supported that the structure of groupwork in POGIL was among the most effective in their undergraduate courses;
4. At the end of the semester, the ACS Semester II Special Exam, which contains paired traditional and conceptual questions on the same topics, was administered. POGIL students achieved significantly higher scores on both components than a stratified random sample of non-POGIL students.

Not all POGIL Classes Are Identical

Even though both sections of POGIL involved in this research were initially similar in characterizations, such as logical reasoning and previous POGIL experiences, a few significant differences between the two sections were observed and analyzed.

Achievement measures through the ACS Second Semester Special Exam found that students in section B achieved higher conceptual scores than students in section A. There had been no significant difference in logical reasoning ability as measured at the beginning of the study.

Additionally, a statistical interaction effect was observed, which identified non-parallel trends in students GALT level and the ACS-traditional subscores. Low GALT students in section A achieved significantly higher traditional scores than the low GALT students in section B. Medium and high GALT students in section A had lower ACS-traditional scores than students in section B.

To explain this observation, more in-depth analysis of the implementation of POGIL in the sections was made. Differences were observed in:

1. Form of the mentor bridge – Professor A often responded to questions from groups with direct answers. Professor B responded with additional questions in guiding the group to an answer.
2. Teacher-prepared tests – An analysis of each professor's one-hour tests by two different chemical educators revealed that content and level of these tests were in the same range. Professor B, though, specifically asked students to explain the reasoning for their responses. Professor A did not explicitly ask students to provide such an explanation. For example, on tests covering pH, professor A asked: Circle the one correct answer in each of the following: Which of the following 0.1 *M* solution has the highest pH? a) KCl; b) HF; c) NaCN; d) NH_4Cl; e) H_2SO_4. Professor B asked: For the following question, select the best answer and explain your reasoning. No credit will be given for an answer without an explanation: Which of these produces a solution with the highest pH?a) x *M* NH_3; b) x *M* PH_3; c) x *M* NH_4Cl; e) impossible to tell without further information.
3. Time spent in particular phases – Differences were observed between the two sections in percentage of time spent in Phase I, Phase II, and Phase IV, as well as in the use of the Tutoring Bridge. Students in section A spent more time checking answers (Phase I) than students in section B, but spent less time discussing discrepancies in their answers (Phase II). Even though both sections spent equivalent time in Phase III, students in section B, based on videotape analysis, had more thorough reviews of their responses. The extended amount of time building their ideas (Phase II) was viewed as a

significant reason for the higher quality Phase III interactions in section B. The combination of more extended Phase II interactions and higher quality Phase III interactions made it more likely for students to be confident in their understanding, and thus, more Phase IV interactions in section B. This also made it easier and more frequent in section B for a struggling student to identify a willing tutor (tutoring bridge). Finally, even though there was no difference in the percentage of mentor bridge interactions, the direct answering style of Professor A was viewed as a reason for students not having to continue their Phase II discussions while the questioning style of Professor B was noted as a significant reason for students' more extended Phase II interactions and more enriched Phase III interactions.

4. Form of the recorder's report – Professor A requested that groups transcribe their responses to questions in the ChemActivity as their recorder's report. Professor B requested that students describe the most important points and summarize the day's lesson. This difference in processing is analogous to the difference in a Phase I interaction and a Phase III interaction.

These observations have implications for the POGIL instructor. Students' logical reasoning ability should be considered when responding to questions. For example, in traditional, algorithmic content, students with low reasoning ability benefit more from direct answers from the professor. When developing conceptual understanding, students of all ability ranges improve when guided by more questions to extend their discussions. Following discussions, students should often confirm their reasoning by sharing ideas not only with each other but also with other groups or the whole class. Even when asked similar questions, students who must provide reasoning for their responses appear to develop more lasting conceptual understanding. Finally, in summarizing lessons, asking students to synthesize the key ideas as opposed to rewriting their responses to questions appears to deepen students' conceptual understanding.

Summary

Students experience Phases and Bridges of learning as they interact during POGIL instruction. Characteristics of these Phases and Bridges are consistent with theories of learning, such as constructivism and the learning cycle. Students have validated these Phases and Bridges and commented that they found them helpful to their learning of general chemistry principles. Finally, small variations in POGIL implementation can cause significant differences in outcomes, such as conceptual understanding, for students in POGIL classrooms.

References

1. Tudge, J. R. H.; Winteroff, P.A. *Human Development.* **1993**, *36*, 61-81.
2. von Glasersfeld, E. *Questions and Answers about Radical Constructivism.* In *The Practice of Constructivism in Science Education*; Tobin, K., Ed.; Lawrence Erlbaum Associates: Hillstock, NJ, 1993.
3. Johnson, D. W.; Johnson, R.T. *Cooperation and Competition: Theory and Research*; Interaction Book Company: Edina, MN, 1989.
4. Lawson, A.E. *Science Teaching and the Development of Thinking*; Wadsworth Publishing Company: Boston, MA, 1994; pp 134-139.
5. Spencer, J.N. *J.Chem.Educ.* **1999**, *76*(4), 566-569.
6. Farrell, J. J.; Moog, R.S.; Spencer, J.N. *J.Chem.Educ.* **1999**, *76*, 570-573.
7. Springer, L.; Stanne, M.E.; Donovan, S.S. Effects of Small-Group Learning on Undergraduates in Science, Mathematics, Engineering, and Technology: A Meta-Analysis, Research Monograph No. 11. *National Institute for Science Education*, 1994.
8. Daubenmire, P.L.; Bunce, D.M. *J. Res. Sci. Teach.,* in press.
9. Roadrangka, V.; Yeany, R.H.; Padilla, M.J.. *GALT. Group Assessment of Logical Thinking.* University of Georgia, Athens, GA. 1982.
10. Gay, L. R.; Airasian, P. *Educational Research: Competencies for Analysis and Application , 6th ed;* Prentice-Hall, Inc.: Columbus, OH, 2000; p 177.

Chapter 9

A Theory-Based Evaluation of POGIL Workshops: Providing a Clearer Picture of POGIL Adoption

Diane M. Bunce, Katherine Havanki, and Jessica R. VandenPlas

Chemistry Department, The Catholic University of America, 620 Michigan Avenue NE, Washington, DC 20064

Typical evaluations of workshops consist of asking participants to rate a workshop's "effectiveness". The questions are written by the workshop organizers and the evaluation is often reported as simple percentages of how many participants liked the workshop. By contrast, in this theory-based evaluation, a more precise instrument based upon the theories of Rogers and Ajzen was developed to determine the participants' stage of adoption readiness, a measure of their intention to adopt POGIL following the workshop, and the type of support they needed. Analysis of these surveys addressed issues such as who attends POGIL workshops, why they come, their stage of adoption readiness, and differences in the barriers and support needed by participants at different stages. Such specific information as a result of surveys was used to positively impact the structure of future workshops and enhance implementation efforts.

The time-honored tradition of presenting in-depth information to other professionals regarding teaching innovations via workshops has been used by most national teaching projects. These workshops vary in length, but typically last one to three days and provide the participants with hands-on experience of the teaching innovation. Such workshops include an overview, opportunities for the participants to experience the teaching innovation and discussion on how they might use it in their own classrooms. Organizers and funding agencies encourage evaluation of the effectiveness of the workshops. This is often accomplished by surveying the workshop participants during the last hour of the workshop, asking if the participants found the workshop worthwhile or effective, and sometimes asking what additional information the participants need. Results of such surveys are usually confined to percentages of participants who rated the workshops on a Likert scale survey from "very effective" to "not very effective". Up to now, this has been accepted as evidence that the workshop did or did not meet the needs of the participants.

In the POGIL project, we were interested in a more targeted evaluation that would help us determine the stage of readiness of our participants to adopt POGIL. We hypothesized that some participants were attending the workshops to attain an overall understanding of what POGIL is and thus were at a beginning stage in the adoption readiness process, while others had tried POGIL-like approaches in their classes and were interested in learning more specific information about implementation. Participant groups at each of these stages of readiness to adopt POGIL would be interested in different things in the workshops. To make the POGIL workshops effective, we wanted to better match our workshop goals to those of our audience.

The primary purpose in organizing the POGIL workshops was to provide pertinent information to the participants so that they would be empowered to adopt POGIL if they so chose. In order to use the evaluation to gain specific information, we employed Rogers' Theory of Diffusion of Innovations (*1*), which provided a way to assess the progress of participants in the innovation-adoption process. A second theory, Ajzen's Theory of Planned Behavior (*2*), provided a framework for structuring questions to address the three components of a participant's intention to choose a behavior. The combination of these two theories provided us with the theoretical framework necessary to construct both a pre- and post-workshop survey that categorized a participant's stage of readiness to adopt and identified participants' beliefs about attitude, perception of peer pressure, and barriers to implementation that comprise their overall intention to adopt this innovation (POGIL).

Theory

In his Theory of Diffusion of Innovations (*1*), Rogers describes an *innovation-decision* model comprising stages through which a participant progresses in deciding to adopt an innovation. Rogers' theory describes five stages in the process. We modified Rogers's innovation-decision stages to correspond to the specific process that POGIL workshop participants would be experiencing. These modified adoption-readiness stages included two additional stages in the model, namely, Stage 1: Dissatisfaction with current teaching practices and Stage 6: Expert--already implementing POGIL and willing to share expertise as workshop leaders. Stage 6 was added to the POGIL model to categorize the contribution of the POGIL Principal Investigators who had already been teaching POGIL for a number of years. A comparison of Rogers' Adoption Readiness Stages and the modified POGIL Adoption Readiness Stages are listed in Table I.

It was determined that participants in the POGIL workshops would be at least Stage 2 in the *adoption-readiness* model by their action of registering for the workshop. Since Stage 6 Experts were not registering as participants in the workshop, but rather acting as workshop leaders, and thus not required to complete the survey, the survey was written to differentiate between Stages 2 through 5 in this model.

Identifying participants at different stages of *adoption readiness* was not enough to predict their intention to adopt POGIL. Ajzen's Theory of Planned Behavior (*2*) was used to analyze the participants' intention at each of the stages of *adoption readiness* (Stage 2—5) to adopt POGIL. Ajzen's theory deals with intention and the factors that affect behavior. The underlying assumption is that the stronger a subject's intention is to do something, the more likely this intention, barring any unforeseeable outside constraints, will result in a specific behavior. Ajzen identifies three components of planned behavior, namely:

- Attitude (Subject's opinion).
- Subjective Norm (Peer pressure to demonstrate or not demonstrate a behavior.)
- Perceived Behavioral Control (Subject's perception of the difficulty (Barriers) or ease in demonstrating the behavior.)

Each of these three components can be measured through a directed series of questions that Ajzen has outlined (*2*).

The combination of the POGIL-modified Model of Adoption-Readiness and the Ajzen Theory of Planned Behavior shaped two surveys (pre and post workshop) that were administered to POGIL workshop participants. The analysis of these two surveys enabled us to address targeted questions of adoption readiness stage and the barriers and support that participants at each stage needed to successfully implement POGIL.

Survey Development

In order to ensure that the survey instruments would be valid, a preliminary open-ended survey was administered to participants during the first year of the POGIL project. Participants were asked to 1) respond to questions regarding their view of the perceived advantages/disadvantages of implementing POGIL;

Table I. Comparison of Rogers' and POGIL-modified stages

Stage	Rogers' Innovation Decision Model	POGIL Adoption Readiness Model
1	**Knowledge**	**Dissatisfaction** with current teaching practices
2	**Persuasion** Participant forms an opinion about the innovation.	**Curious** Comparable to Rogers' Stage 1: Knowledge.
3	**Decision** Participant decides to either adopt or reject the innovation.	**Willing** Comparable to Rogers' Stage 2: Persuasion.
4	**Implementation** Participant adapts the innovation to his/her specific situation and implements it.	**Planning** Comparable to Rogers' Stage 3: Decision.
5	**Confirmation** Participant seeks verification of his/her decision.	**Implementing** Combination of Rogers' Stage 4: Implementation and Stage 5: Confirmation.
6		**Expert** Willing to share expertise with others.

2) identify people who supported or did not support their implementation of POGIL; and 3) list factors that affected the ease of implementation of POGIL at their home institutions. The 106 participant responses from this open-ended survey were used to construct a 7-point Likert scale instrument that was used with the larger population.

The Likert scale presurvey was administered to workshop participants as part of their online workshop registration. Data were collected using a commercial web service (*3*). Workshop participants were invited to complete an online post-workshop survey during one or two semesters following the completion of a workshop. This time frame was chosen to allow participants an opportunity to try POGIL in their classes if they so chose before completing a post workshop survey.

Participants were categorized according to the adoption readiness stages on the basis of five questions on the presurvey and seven questions on the post survey. These questions included topics such as:

- Is this the first POGIL workshop the participant had attended?
- The reasons selected for attending the workshop.

- Did the participant currently use POGIL?
- An indication of how often POGIL was used in class.
- The methods participants used to learn about POGIL.
- The amount of time spent in the learning process.
- The participant's intention to use POGIL in the next two semesters.

Ajzen's theory was represented by specific sections of the survey that probed for attitude, subjective norm (peer pressure), and perceived behavioral control (barriers).

The analysis of the survey data included 1201 complete presurveys from participants who attended one of 51 different workshops over two and a half years of the Project. The response rate for the post-workshop survey was 27% with 329 participants completing online post surveys over the same time period. Of the people who completed both pre and post workshop surveys, 199 matched sets of data were analyzed. Although this is not a very high response rate, it is not atypical (*4*) for national surveys.

Questions Addressed by the Data

The use of surveys incorporating Rogers' (modified) and Ajzen's theories enabled us to address specific questions that are of importance for an in-depth evaluation of the effectiveness of the POGIL workshops. These questions include the following:

- Who attends POGIL workshops? Is there a difference in who attends the one-day vs. the three-day workshops?
- Why do participants attend POGIL workshops? Is there a difference between one vs. three-day workshop participants' reasons for attending?
  Do participants at different stages of adoption readiness have different reasons for attending?
- Do participants at different adoption readiness stages:
  - demonstrate different attitudes towards POGIL?
  - identify different sources of peer pressure in the adoption process?
  - identify different barriers to implementation of POGIL?
  - indicate different kinds of support?
  - identify different types of additional help needed to support their implementation?
- Do people change adoption readiness stage from pre- to post-workshop surveys?

Who Attended POGIL Workshops?

Because there were more one-day than three-day events during the period studied, most POGIL workshop participants (63%) attended one-day rather than three-day workshops. The single most common home institution category of the participants was a 4-year undergraduate college (35%). The single largest category for length of time participants had taught was more than 10 years (42%) and the most commonly reported teaching status was tenured (42%). The most common content area taught by workshop participants was chemistry (79%). Not surprisingly, given that the period under study was soon after the establishment of the POGIL project, the majority of participants were attending their first POGIL workshop (76%). The distribution of stages among participants (n= 1201), as determined by the presurvey, is shown in Figure 1. Somewhat surprisingly, of the people attending the POGIL workshops, the majority (56%) reported that they had already implemented POGIL (Stage 5), while 30% reported that they are planning to implement (Stage 4).

Why Do Participants Attend POGIL Workshops?

Analysis of the reasons why participants in general attended the workshops showed that roughly one third of the participants (36%) did not specify any particular reason for attending the POGIL workshop while another third (33%) came because they wanted more general information regarding POGIL.

However, participants at higher stages of adoption readiness give reasons for attending a POGIL workshop that are increasingly more specific and geared towards the details of implementation. The results of the survey on this point are given in Figure 2. These percentages reflect the relative importance assigned to each question by participants within a given stage. The need for general information about POGIL becomes less important at the higher stages (Stage 3 (53%); Stage 4 (40%) and Stage 5 (28%)). The need for answers to specific questions about implementing POGIL is highest for those participants who are planning to implement (Stage 4 (21%)) followed by those who are implementing (Stage 5 (14%)). Stage 5 participants also want to gain information at the workshops that will improve their implementation (26%). All Stage 6 participants attend the workshops because they want to share their expertise (100%).

Although many participants select one-day vs. three-day workshops based primarily upon the workshop's proximity to the participant's home institution or the date of the event, the survey data suggest that some participants select a three-day workshop after attending a one-day workshop. When the responses from one-day workshop participants are compared to those of three-day workshop participants, 83% of one-day workshop participants report that they

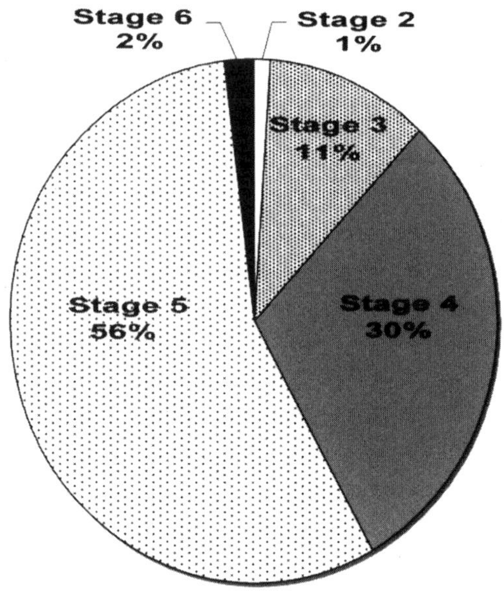

Figure 1. Distribution of participants by stage

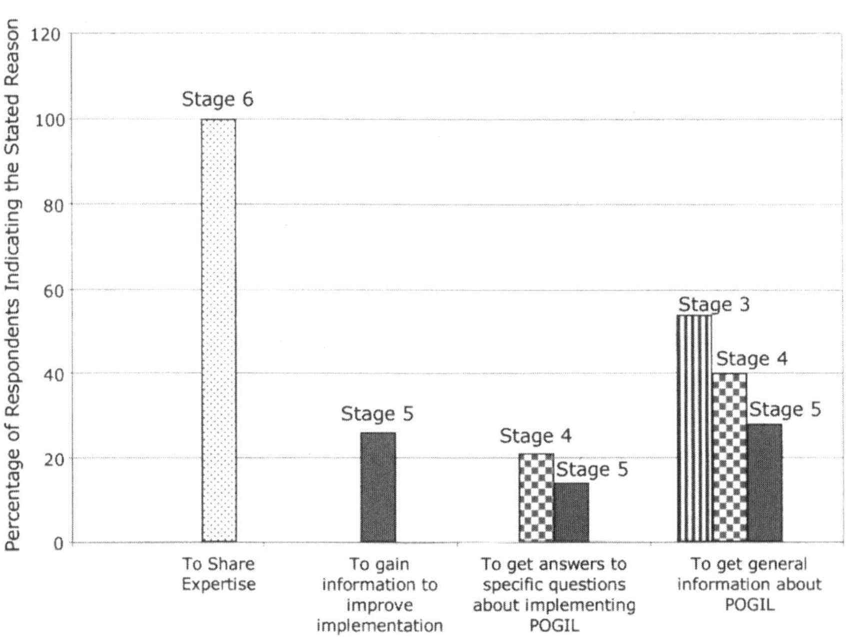

Figure 2. Reasons for attending workshop by stage

are attending their first workshop while only 64% of three-day workshop participants are attending their first workshop. This means that it is not uncommon for people who attend a one-day workshop to then attend a three-day POGIL workshop for additional information and experience with POGIL. This is further corroborated by the fact that the percentage of people at Stage 5 (Implementing) is higher at three-day workshops (64%) than one-day workshops (52%). The interpretation of this trend is that three-day workshop participants are looking for more specific information dealing with their implementation of POGIL than one-day workshop participants.

Do Participants at Different Stages Differ in their Attitude, Response to Peer Pressure or Identification of Barriers to Implementing POGIL?

Categorization of POGIL workshop participants by stages of adoption readiness enables us to analyze the differences in these participants' perception of the three components of Ajzen's Theory of Planned Behavior (Attitude, Subjective Norm (Peer Pressure), and Planned Behavioral Control (Barriers)).

Attitude

Participants' attitude towards POGIL was determined by their answers to three questions on the Presurvey. These questions asked the participants to rate their responses to questions about the implementation of POGIL on the effectiveness of their teaching; benefit to students; and overall importance. The ranking consisted of a 7-point Likert scale with 1=Extremely Detrimental (Unimportant) to 7= Extremely Beneficial (Important).

Participants in Stages 4 (Planning), 5 (Implementing), and 6 (Expert) ranked each of these questions significantly higher (Ave. = 5.97) than participants in Stages 2 and 3 (Ave. = 4.42). This is viewed as evidence that participants in higher stages of adoption readiness demonstrate a more positive attitude towards the innovation as evidenced by their higher rating of its effectiveness and importance and thus are more likely to actually implement POGIL than participants characterized as Stages 2 (Curious) or 3 (Willing).

Subjective Norm (Peer Pressure)

Questions on the presurvey asked about the participants' perception of support from their peers for the implementation of POGIL and adoption of innovations in general. These questions used a different 7-point Likert scale that ranged from 1= Extremely Likely to 7= Extremely Unlikely to demonstrate support.

Participants at the higher stages of adoption readiness report relatively high levels (Quite or Extremely Likely) of perceived support from their peers for the adoption of POGIL (Stage 4: 67%; Stage 5: 69%; and Stage 6: 66%). When this trend is explored in more detail, it becomes obvious that participants in Stages 4-6 perceive their fellow faculty (both tenured and untenured), Administration, Professional Colleagues, and students to likely support their adoption of POGIL. Stage 3 (Willing) participants followed the same trend as Stages 4, 5, and 6 participants but rated each of the peer support groups lower by 10 to 20% on the combined Quite and Extremely Important score.

Support staff and teaching assistants are perceived as supportive of the adoption of POGIL (Ave. = 40%) by the participants' who are in adoption stages (Stages 3-5). Thus the data are interpreted as peer support from faculty and administration at a participants' home institution along with that of professional colleagues and students (67% average across stages) are the groups that active POGIL adopter participants (Stage 4-6) expect will most likely support their adoption of POGIL, while the support of staff and teaching assistants for the adoption is not perceived as being as likely.

Perceived Behavioral Control (Barriers)

The participants' categorization of barriers to implementation of POGIL was determined by the factor analysis of their ratings of 18 specific barriers to implementation on a 7-point Likert scale of 1 (Extremely Unimportant) to 7 (Extremely Important). The 18 barriers included a range from the amount of time to implement POGIL to budgetary concerns. The complete list of barriers can be found in Table II. All participants, regardless of stage, rated barriers such as instructor knowledge of POGIL techniques, availability of POGIL materials, and access to POGIL experts (barriers c-f in Table II) as important.

Further analysis revealed a difference in the pattern of remaining barriers rated by participants in different stages of adoption readiness (Table II).

Participants in Stage 2 rate almost all barriers (17 out of 18 barriers) as important to their decision to adopt POGIL. Stage 3 and Stage 4 participants rate a reduced number of barriers as key (12 and 13 barriers respectively). Stage 3 participants do not rank either of the two barriers concerned with time to implement or sustain POGIL as important, while Stage 4 participants do see these barriers as important. Stage 3 participants do not rate barriers that are out of their direct control such as access to appropriate class facilities, technology and scheduling as important, where Stage 4 participants do. Another difference between the participants at these two stages is that Stage 3 participants are concerned with the presence and training of TA's while Stage 4 participants are not. This difference may be due to the different decisions that Stage 3 (Willing)

Table II. Factor Analysis of Perceived Behavioral Control (barriers) by Adoption Readiness stage

Question from Survey	Stage 2	Stage 3	Stage 4	Stage 5
a. the amount of time to implement POGIL	X		X	
b. the amount of time required to sustain POGIL	X		X	
c. instructor knowledge of POGIL techniques	X	X	X	X
d. POGIL materials for my course	X	X	X	X
e. access to other successful POGIL implementers	X	X	X	X
f. access to POGIL experts	X	X	X	X
g. multiple teachers for the same course at my school	X	X		X
h. the presence of teaching assistants	X	X		X
i. training of teaching assistants		X		X
j. budgetary concerns	X	X	X	X
k. class size	X	X	X	
l. student evaluations of course	X	X	X	
m. tenure and promotion	X	X		
n. current teaching practices in my school	X	X		
o. access to appropriate classroom/lab facilities	X		X	
p. access to adequate technology	X		X	
q. the amount of freedom I have in my department to implement innovation	X		X	
r. scheduling of classes (recitation, class length, etc.)	X		X	

and Stage 4 (Planning) participants must deal with to implement POGIL. Stage 3 participants may have already come to the conclusion that it will take time to implement POGIL and they now are turning their attention to the other barriers that might interfere with their implementation such as tenure/promotion issues and teaching assistants. By Stage 4, the participants have probably already dealt with the potential problems within the department such as student evaluations and tenure/promotion issues and are now concerned with the practical issues of class facilities and scheduling. Stage 4 participants also turn back to the issues of the time it is going to take to implement POGIL now that they are heavily into the planning stage.

Stage 5 (Implementing) participants list the fewest number of barriers (10 of 18) but still include knowledge of POGIL, availability of POGIL materials and access to POGIL implementers and experts among important barriers. Stage 5 participants do not rate issues of student evaluation or tenure/promotion as important as participants in Stages 2 and 3 do. This may be due to the groundwork that Stage 5 participants have already done to keep the department informed of their implementation of POGIL.

The analysis of this targeted survey points out the differences in perceived barriers that occur in workshop participants who are at different stages of adoption readiness. Based on this data, workshops were revised to include more emphasis on dealing with the barriers that each stage might see as important.

Summary of Planned Behavior Components by Stage

The results from the presurvey show that participants in Stages 4 and 5 are more positive about the perceived benefits of implementing POGIL, expect support from their peers, and shift the selection of barriers they face during an implementation of POGIL from an all-encompassing laundry list to a more specified list. Stage 5 participants, who have either overcome or decided to discount a large number of barriers, focus on the barriers that are in their immediate control. This information was helpful in planning POGIL workshops during the project because the specific concerns of those who were implementing (Stage 5) and planning to implement (Stage 4) could be included in specific features of the workshop. The workshops were modified to lend support to those who needed specific information about POGIL and helped address specific barriers identified by participants including networking with current POGIL implementers and increased access to POGIL leaders. Workshops were also sometimes organized with parallel sessions where participants at lower stages of adoption readiness could attend a session that addressed the general issues connected with POGIL while other participants attended sessions that addressed more specific concerns.

Do Participants at Different Stages Identify Different Kinds of Support as Necessary for POGIL Implementation?

Questions were included on the post survey that addressed the type of support participants chose as important for their implementations of POGIL. Analysis of the data shows that participants in Stage 3 (Willing), Stage 4 (Planning) and Stage 5 (Implementing) all chose the following as important types of support for implementation of POGIL:

- contact with a colleague using POGIL
- participating in a consultancy (either a POGIL leader visits your institution or you visit one of the POGIL leaders' institutions)
- observing a colleague who is teaching POGIL
- contact with a POGIL leader

In addition, analysis of participants who were currently implementing POGIL and those who had previously implemented POGIL but were not currently doing so showed that these two groups chose attending a three-day workshop as an important support for implementation.

What Other Types of Help Do Implementers Want?

Participants who identified themselves as implementing POGIL and those who had implemented previously but were not now using POGIL, were asked to select the areas where they would like additional help with implementation of POGIL. Both groups rated the five types of additional help (Evaluation of Student Achievement; Effectiveness of POGIL; Classroom Management; Teaching Techniques; and Development of Additional POGIL materials) as equally important. There were no significant differences in the ratings between the two groups.

These additional areas of help deal with sustained or expanded implementation of POGIL, and verify the idea that support for adoption of innovations must be provided beyond the initial implementation and must address issues arising from continuing implementation.

Do Participants Change Stage of Adoption Readiness from the Pre- to Post-Workshop Surveys?

Only those participants who answered both the pre- and post-surveys could be used in this analysis. There were 199 people who answered both surveys and most of these people (65%) attended one-day workshops. The demographics for this group of matched pre- and post-surveys was similar to that of the overall sample of 1201 who answered the presurvey.

In terms of the relative number of people at each stage of adoption readiness from the pre to the post survey, the following trends from pre and post survey administrations were found:

- 47% of participants stay at the same stage of adoption readiness
- 29% increase one or two stages higher
- 26% decrease one or two stages lower

Further analysis shows that the biggest gain (20%) occurred in participants who moved from Stage 4 (Planning) to Stage 5 (Implementing) from the pre to the post survey. The largest drop (10%) was from Stage 5 (Implementing) to Stage 4 (Planning) and from Stage 5 to Stage 3 (Willing) (9%). This drop was partially offset by the 6% of Stage 3 (Willing) who moved to Stage 5 (Implementing).

The movement of some participants to lower levels of adoption readiness may be a result of the workshops serving as a reality check for what actually constitutes a POGIL implementation. This may cause some participants to re-evaluate their declaration that they are currently implementing POGIL. On the other hand, other participants may have gained the necessary confidence from the workshops to increase their extent of implementation.

Summary

The use of theory-based evaluations can provide a rich data set for the evaluation of participants' needs and effective workshop design. By measuring both the workshop participants' stage of adoption readiness and each stage's attitude, perception of peer pressure and barriers to implementation, workshops can be targeted to address these issues. Use of precise feedback can significantly increase the effectiveness of workshops designed to support change in teaching behavior. Surveys should be viewed as instruments that can be designed to effectively measure the perceptions of respondents. Development of such tools requires investing time to develop them in accordance with theories of change and validating them by using open-ended responses to meaningfully determine the selection of options to be rated by the population of interest.

In the case of the POGIL workshops, these theory-based surveys were able to track the differences in participants' attitude, perception of peer pressure and identification of barriers to implementation in different stages of adoption readiness. The result was a project that could more effectively respond to the needs of its participants both within and outside of the workshop format. The goals of the POGIL project were many but the project broke new ground on the content of the innovation; the way the innovation was presented to participants; the development of tools to measure the stage of adoption readiness; and perceptions of participants at each stage.

References

1. Rogers, E. M. *Diffusion of Innovations*; 4th ed.; The Free Press: New York, 1995.
2. Ajzen, I.; Fishbein, M. *Understanding Attitudes and Predicting Social Behavior*; Prentice Hall, Inc.: Upper Saddle River, 1980.
3. *Formsite.* URL http://formsite.com. Last accessed October, 2007.
4. *Tercent, Inc. - SuperSurvey.*
   URL https://secure.supersurvey.com/whitepapers.htm.
   Last accessed October, 2007.

Chapter 10

POGIL in the High School Chemistry Classroom

Laura Trout[1], Linda Padwa[2], and David Hanson[3]

[1]Manheim Township High School, Lancaster, PA 17606
[2]Science Education Program, Stony Brook University, Stony Brook, NY 11794
[3]Department of Chemistry, Stony Brook University, Stony Brook, NY 11794

High school teachers are using POGIL materials in their classrooms because POGIL addresses the need for inquiry-based lessons as outlined in the National Science Education Standards. However, several implementation issues need to be addressed in the transition of POGIL from colleges to high schools. These issues include the range of topics covered, the rigor of the materials, and the perceived difficulties associated with the use of cooperative learning in high schools. High school teachers have found that they need to alter college level materials or write new materials in order to effectively use POGIL in their classes. Attending a teacher-training session also enhances the confidence of HS teachers who use POGIL.

Good, innovative teachers are always looking for the "best" way to convey their subject matter to their students. They regularly seek out successful techniques that are used by colleagues in their own school, or elsewhere, and adapt these techniques for use in their classes. Although the POGIL methodology was originally developed for use at the college level, in many ways it is also an excellent model for use in the high school classroom. This chapter addresses several questions and issues related to this adaptation, including how college level materials need to be modified (or rewritten) for use in high school settings (grades 9 – 12) and how implementation strategies may need to be changed for high school classrooms.

A Good Fit

The success of POGIL in first-year college chemistry programs is an indicator that it can be successful when used with high school students; after all, first year college students are only a year or two removed from high school chemistry. The attitudes that new college students bring to their lectures and recitations are not too different from the attitudes of their slightly younger high school friends. However, when first introduced to POGIL, many high school teachers can be skeptical about using the POGIL methodology with their students. Typical comments include: "The students don't have enough background knowledge" or "They are not mature enough" or "I need to cover so much material for standardized testing, I can't afford the time for inquiry learning." Over the past several years, countless high school teachers have proven the skeptics to be wrong.

The National Science Education Standards (*1*) address the expectation that science will be taught with an emphasis on learning through inquiry. Many, if not most, high school teachers agree that inquiry and the use of materials based on the learning cycle are the best methods for teaching. However, many of the materials currently used in high schools do not place sufficient emphasis on these aspects of instruction. Even if teachers are knowledgeable about teaching with process-centered techniques, they often find that existing curriculum materials do not adequately support inquiry-based learning in their classes. Since POGIL is rooted in scientific process, guided inquiry, and constructivism (where students create their own knowledge based on experiences), it is actually a perfect fit for use in high schools. POGIL materials provide students with a solid foundation of scientific thought processes and content.

However, there are some differences between the high school and the college or university setting. Experience has shown that unlike in the college setting, POGIL materials should not be used as the only method of instruction in high school classes. Furthermore, currently available college materials need to be modified, or new materials need to be written for the younger audience. Finally, process skills may need to be modeled more explicitly for high school students than for college students.

Why Not Use College Materials in High School Classes?

Many high school teachers have found that the POGIL materials developed for use in college-level general chemistry courses (*2, 3*) can be used successfully in AP chemistry classes. However, a number of issues have been identified concerning the use of these materials more generally in high school courses:

- College materials do not necessarily address the Chemistry Content Standards set by the National Science Education Standards, nor the science/chemistry standards set by individual states, e.g., New York (*4*).
- College materials are designed for use in class sessions that are longer than the typical high school class.
- In the college materials concepts are developed in "chunks" that are too big for high school students.
- The vocabulary and math level in some college materials are not appropriate for high school students.
- The questions in college materials are written so that students are often challenged to combine newly learned skills with prior knowledge in a way that would be too demanding for high school students. (Keep in mind that high school students do not yet have this prior knowledge.)

Many of the POGIL lesson materials currently used in high school courses are modified versions of materials that were originally written for use in college courses. These were written by high school teachers and are available through the POGIL website (*5*). In many cases the knowledge level for high school students is not of equal depth, or the content being emphasized is different, when compared to college materials. For example, a college chemistry course may not devote a large amount of time to teaching students about measurement and significant figures, but this is a recurring concept in a high school chemistry course. Since some high school students may have difficulty making the large mental jumps that are typically required between questions as written in the college materials, more exploration questions are needed in order for them to become comfortable with the model(s) provided in each lesson, and more guidance is needed for them to fully develop the concepts being taught. [The *model* in a POGIL lesson (or activity) contains information that students use as they answer questions and construct their own understanding about the concept being addressed in the activity.]

In some instances, part, or all, of the model in an activity can be constructed by the students. For example, in an activity developed by one of the authors (LT) on atomic structure, students use beads of different colors and sizes to represent protons, neutrons and electrons in atoms. The students construct models of atoms by placing the beads in sealed plastic baggies using specific instructions on how many proton, neutron or electron beads to place in each bag. During the next lesson the bags are used to work through a POGIL activity that introduces atomic number, mass number, and the concept of isotopes.

When the model is more mathematical, students may be asked some exploration questions that guide them as they figure out how numbers in a data table are obtained. Once they have this understanding, students can be guided to apply their new knowledge to complete data tables that have missing pieces of information.

Students may need to be directed more deliberately when asked to determine the relationship between data points presented in a model. One effective strategy to help them in this development is the use of language in the guiding questions that is similar to language that is encountered in their math classes. Having them reconstruct a table with the data in a certain order, or having them draw a graph and describe the relationship, will help students to see patterns in the data more readily.

Implementation Issues

As many teachers know, high school students will get bored with any method used fanatically. If there is lecture all the time they get bored; if there are labs all the time, they get bored; and if there are too many POGIL activities in a row, they get bored. When students are bored, any instruction loses its educational value and it takes much teacher ingenuity to keep them engaged. Therefore, POGIL activities should be one of many different techniques used by teachers while planning a unit. Implementing at least one POGIL activity every week or two, particularly at the beginning of a lesson or unit to introduce key concepts, can be an effective approach.

The group interactions associated with the POGIL methodology can also present challenges when working with high school students. These students tend to be more diverse in abilities than students found in college and university courses. Some reasons for having students work in cooperative groups include developing good communication skills and working together to reach a higher level of understanding by asking questions and explaining concepts in their own words. Many students, however, think they are working in a group to get finished with the activity faster. They frequently don't want to help the weaker students understand the concepts, but if group structure requires it, they will simply tell these students what answer to write on their paper. Therefore, it is necessary to be very deliberate and explicit about the instructor's expectations during group interactions.

At the beginning of the year, before doing a POGIL lesson, it is important to discuss what it means to work as a group. Students can be asked to share situations with the class that illustrate a group that works well together, and situations that illustrate a dysfunctional group. By staging role-playing situations, such as a "strong" student in the group giving away the answers, or a group where there is appropriate group interaction leading to full understanding by all members, students can be introduced to ineffective and effective group behaviors. These scripted role-plays are effective when discussed with the whole class. During the first few POGIL lessons, and throughout the year, students can be reminded of these discussions and they can be asked to reflect on the group interactions they have experienced. When asked about the behavior of their own group, students can be very insightful and suggest excellent approaches that would improve their group's performance.

From the instructor's perspective it is important to reward good group interaction, and not personal performance. For example, individual POGIL activities should not be graded for correctness, but one paper might be chosen at random to represent the group's work. Rather than rewarding groups for completing their work quickly, instructors can call attention to groups that have rousing discussions amongst members as they answer questions. While circulating throughout the room, the teacher can direct discussion to the weaker members of a group to make sure that they fully understand the new material. When a group quiz is given to measure understanding, some method needs to be devised to ensure that all members participate, rather than have one student answer all the questions while the others watch. For example, a quiz may contain three questions, with each student in the group directed to answer one of the questions. Group members can discuss their own strengths and weaknesses and distribute the questions accordingly, so each member gets to show his or her strength, giving the entire group a better score. Whatever is done to "grade" the POGIL activity, it is important to validate the *group* work, and not simply the performance of any individual. Students are very good at discovering what the instructor really values, so sending the correct message is vital.

Many high school instructors have found that groups of three students are ideal. Two does not allow for enough interaction, and four creates too much potential for "chit-chat". Some instructors have found that making the groups heterogeneous by achievement is not optimal since high school students tend to know each other very well, and they can readily identify when the groups have been crafted to have a "low", "middle" and "high" achiever. Thus, randomly assigning groups is a good alternative. Typical roles used in three-person groups are Manager, Reader, and Presenter. The Manager is the only one who may ask the instructor a question and is responsible for making sure all members of the group are confident about the material before moving on. The Reader reads the instructions and questions aloud to the group. This allows slow readers or English Language Learners to participate fully without embarrassment. It also keeps groups together on the same question, or at least allows the instructor to quickly spot groups that are not working together. The Presenter is responsible for sharing the group's findings during whole class discussions. At the end of each POGIL lesson the Presenters are asked questions about the major content that should have been learned from the activity. Although one of the principles of POGIL is that students learn the content without teacher-directed instruction, in many cases high school students do not have sufficient confidence in what they have learned until it has been confirmed by the teacher. These end-of-activity debriefing sessions are a way to build confidence in students without direct teacher instruction. This is very necessary at the high school level.

Lessons Learned

There are an ever-growing number of high school teachers implementing POGIL in their classrooms. A recent survey by the POGIL Project of practitioners across the country indicated that hundreds of teachers have already implemented some POGIL activities during the 2006-7 academic year and/or the two preceding years. In addition, several initiatives to provide training for teachers who wish to author activities for use in their classrooms have been undertaken, including summer workshops at Arcadia University in association with the Math Science Partnership of Greater Philadelphia and at Stony Brook University. From these experiences, a number of important lessons have been learned:

- Even though teachers may be familiar with the Learning Cycle (exploration, concept development, concept application), and often use it as they plan their lessons, they sometimes omit elements in the execution of their plan. This omission leads to poorer student understanding and less mastery of the material. For example, rather than guiding students in an exploration and leading them to a discovery, they jump in too quickly to provide an explanation.
- Some high school teachers have difficulty facilitating cooperative learning or learning-team methodologies. While the fundamentals of cooperative learning are part of most teacher training programs, not every teacher is comfortable implementing cooperative learning in the classroom.
- Some students are unwilling to be active learners and find themselves frustrated and unsuccessful when working with POGIL activities.
- Teachers who choose to work with the POGIL methodology sometimes have to overcome resistance from colleagues who don't want to change their method of teaching, and who claim that inquiry-based lessons are too difficult for them to implement, and that the activities are too hard for their students to understand.
- Some teachers believe that the best way to prepare their students for college chemistry is to teach by the lecture method.

Based on these observations, some insights have been gained into what might be necessary in order to have even greater success with the implementation of POGIL in high schools.

- Teachers need to have the opportunity to observe POGIL working in a classroom. More teachers will become effective POGIL practitioners after working with well-designed materials and after gaining confidence with the role of group facilitator.
- Teachers need support from project personnel as they introduce POGIL in their classes. There may be a tendency to abandon the methodology if it doesn't go smoothly the first few times it is tried.
- High school students need variety in the mode of instruction that is used each day. POGIL activities can be used once or twice during a unit, but would not be effective as a daily mode of instruction.
- More materials are needed. As previously mentioned, an edited collection of high school POGIL activities is available on the POGIL web site (5). Teachers from around the country have begun to use them, and several teachers have submitted their own lessons that have been added to the growing library of successful materials.
- A Teachers Guide to accompany the high school activities that are on the POGIL web site has been prepared, and is available to teachers who request access to the files. The Guide contains references to the National Science Education Standards, teaching suggestions, and answer keys that can be used to guide teachers as they respond to and grade student work.

Conclusion

The POGIL approach can generate great success with difficult topics in high school chemistry classes at basic, regular and honors levels. Student response to the materials is generally positive. They report that the activities guided them through the development of concepts in a manner that resulted in a better understanding of the material and led to more confidence in their ability to answer questions about the topics. Students using POGIL materials tend to understand concepts better, and retain the understanding longer than with traditional methods. They leave their chemistry classes with skills that extend beyond the acquisition of scientific concept knowledge, such as learning to work as a member of a group, how to organize information, and how to find patterns in data presented in a model.

Plans are being developed to extend the application of the POGIL methodology to additional high school subjects, such as biology, earth science, and physics. As teachers become familiar with POGIL, it won't take long for it to become another effective tool used to improve learning in all of the sciences.

References

1. National Committee on Science Education Standards and Assessment; National Research Council. *National Science Education Standards;* The National Academies Press: Washington, DC, 1996.
2. Hanson, D. *Foundations of Chemistry: Applying POGIL Principles;* Pacific Crest: Lisle, IL, 2006.
3. Moog, R.; Farrell, J. *Chemistry: A Guided Inquiry*; John Wiley and Sons: New York, 2006.
4. *Physical Setting Chemistry/Core Curriculum,* The University of New York State, NY, 2001.
   URL http://www.emsc.nysed.gov/ciai/mst/pub/chemist.pdf.
   Last accessed October, 2007.
5. POGIL website for high school activities.
   URL http://www.pogil.org/materials/high_school.php.
   Last accessed October, 2007.

Chapter 11

POGIL in the General, Organic, and Biological Chemistry Course

Michael P. Garoutte

Physical Science Department, Missouri Southern State University, Joplin, MO 64801

The general, organic, and biological chemistry course or course sequence is, generally intended primarily for nursing and other health science majors. Unique challenges regarding the implementation of the POGIL method in this course are addressed.

The POGIL method has been shown to be effective—in courses for chemistry majors and at various types of institutions—in imparting thorough content knowledge and accompanying process skills to students while reducing student attrition rates (*1-5*). Published, tested materials following this model are available for general chemistry (*6-7*), physical chemistry (*8-9*), and organic chemistry (*10*). However, the use of the POGIL method in an allied health chemistry, or "GOB" (general, organic, and biological chemistry) course presents unique challenges.

Challenges for Allied Health Chemistry

Nationwide, versions of the GOB course range from half-year to full-year offerings. Typically, the course has no chemistry prerequisite, and only a minimal mathematics prerequisite (college algebra). The GOB course or course sequence serves a range of majors—from nursing, dental hygiene, and other health science programs, to medical technology and nutrition—and it is sometimes used even as a core science requirement for some students in programs such as teacher education and psychology.

There is a more or less uniformly accepted set of topics that are considered to be appropriate for the GOB course(s). These topics are not significantly different from those suggested by an ACS task force in the 1980's (*11-12*); however, the topic outline for the two-semester sequence is exhaustive, and takes up four journal pages. Subsets of these topics are normally chosen for a particular GOB course. Commercially available textbooks for the course also contain subsets of the material suggested in the topic outline. There is a smorgasbord of textbooks offering versions of this content at different levels of depth, even to the extent that a single author may have three separate texts with titles containing the words "General, Organic and Biological Chemistry."

While there is general agreement among those who teach such courses that the amount of content is too great for the time allotted, there is no consensus on which topics are most important. One exception to this however, is a budding movement to reduce the amount of traditional organic chemistry in the course, in order to allow more time to focus on the biological chemistry that is relevant to the majority of the clientele, *i.e.,* nursing majors (*13-14*). Indeed, the curricula of dental hygiene programs accredited by the American Dental Association are required to contain content in biochemistry (*15*), and the GOB course is the typical vehicle for this content.

So the challenges of using POGIL in the allied heath class, as compared to a class for chemistry majors, are threefold: first, the students, having previously taken fewer science and mathematics courses, have less preparation; second, the content expected to be covered in the course is excessive; and third, published, tested classroom materials have not until recently been commercially available for the course.

Both the guided-inquiry activities and classroom facilitation techniques for this course must therefore address different issues than for a majors course. A topic that may require a simple definition or brief vocabulary review for a chemistry major may indeed be a concept worthy of a dedicated classroom activity in the allied health course. And, due to the fast pace of the GOB course, some compromises may have to be made in the activities, compared to what would be preferable in a majors course. In this chapter, the particular choices that were made during writing a complete set of POGIL activities for the GOB course (*16*) are discussed.

Topic Selection

The two main challenges to topic selection in the GOB course sequence are the limited mathematics background of the students, and the vast number of topics considered essential to the course.

As mentioned earlier, the mathematics requirement needed to enroll in the GOB course is minimal. Furthermore, many students entering the course may be returning to school after a long absence, with their last math course years in the past. Therefore, these students often have little confidence in their mathematics skills. This leads to the need to dedicate classroom time to developing these skills. For example, in a majors course, an instructor may not need to dedicate much classroom time to developing students' skills at performing simple unit conversions; in the GOB course, this is an essential classroom activity.

Figure 1 contains an excerpt from an activity that explores units and unit conversions using a "unit plan" method—one related to the commonly-employed "factor-label" method. In this activity, the learning cycle (*17*) is employed in the development of students' concepts about quantities, fractions, units, conversion factors, and the process of converting units. An activity of this type is not included in published materials for college general chemistry (*6*), and may perhaps even be considered more appropriate for a high school chemistry course. In fact, the GOB course, at least in the early weeks, may share some similarity to a high-school course. In addition to including this sort of preparatory material, it is a good idea in the GOB course to continually reinforce the mathematics skills developed early in the course by repetition in classroom activities and homework assignments for the entire duration of the term(s).

The other major challenge is the large number of topics that are recommended for inclusion in the GOB course (*11-12*). Lots of content means less time on each topic, and therefore the need for some compromises. Compromise can be made in one of three ways: increasing contact time, credit hours, or number of courses; decreasing the number of topics addressed; or reducing the depth of coverage. Since gaining more contact time is unlikely, and since even nursing instructors and professionals think most of the topics are beneficial (*19-20*), a combination of the last two options seems the best path.

Since, as has been explored, it is not beneficial to eliminate early introductory activities in a nonmajors course, topics considered for omission must come from the "core concepts" of the course. The most likely place for these omissions seems to be in the organic chemistry section—specifically, the multitude of traditional organic reactions that are included in most GOB textbooks. Some instructors have already begun to move in this direction, realizing that the biochemical topics at the end of the textbook are the most pertinent (and interesting) to the students in the course. Only those organic reactions relevant to the biochemical topics to be discussed later in the course need to be included; for example, condensation and hydrolysis reactions are needed to discuss the synthesis and breakdown of triacylglycerols (fats).

With these criteria in mind, a sequence of activities used in a one-semester course might follow that shown in Table I. In this list of 44 activities, the first 27 constitute the general chemistry section; the organic chemistry activities are the next six, and the biological chemistry activities are the final 11.

Table I. List of POGIL Activities (*16*) Completed in a One-Semester General, Organic, and Biological Chemistry Course[a]

1. Working in Groups; Estimation
2. Types of Matter; Chemical and Physical Changes
3. Atoms and The Periodic Table
4. Unit Conversions: Metric System
5. Measurements and Significant Figures
6. Density and Temperature
7. Electron Configuration and The Periodic Table
8. Nuclear Chemistry
9. Ions and Ionic Compounds
10. Covalent and Ionic Bonds
11. Electrolytes, Acids and Bases
12. Naming Binary Molecules, Acids and Bases
13. Molecular Shapes
14. Polar and Nonpolar Covalent Bonds
15. The Mole Concept
16. Balancing Chemical Equations
17. Predicting Binary Reactions
18. Oxidation-Reduction Reactions
19. Mass Relationships (Stoichiometry)
20. Thermochemistry
21. Equilibrium
22. Rates of Reactions
23. Gases
24. Solutions and Molarity
25. Hypotonic and Hypertonic Solutions
26. Acids and Bases
27. Buffers
28. Alkanes, Cycloalkanes and Alkyl Halides
29. Conformers
30. Constitutional and Geometric Isomers
31. Isomers
32. Properties of Organic Molecules
33. Reactions of Organic Molecules
34. Carbohydrates
35. Lipids
36. Amino Acids and Proteins
37. Energy and Metabolism
38. Enzymes
39. Nucleic Acids
40. Glycolyis
41. Citric Acid Cycle
42. Electron Transport/Oxidative Phosphorylation
43. Fatty Acid Oxidation
44. Other Metabolic Pathways

[a] Four practice classroom activities are also included in the course; two on stoichiometry (mole relationships), one on gases, and one on organic functional groups

ChemActivity 4 Unit Conversions: Metric System

Model 1: Fuel efficiency of a particular automobile

- A particular automobile can travel 27 miles per gallon of gasoline used.
- The automobile has a 12-gallon gasoline tank.
- At a particular location, gasoline costs $3.00 (3.00 USD) per gallon.

Critical Thinking Questions:

1. Three statistics are given in Model 1. Circle the two statements that give numerical **ratios**.
2. One statement in Model 1 gives a measured **quantity**. Write the quantity (with the associated **unit**).
3. Write each ratio that you circled in Model 1 as a **fraction**. Your fraction should have a **number** and a **unit** in both the numerator and the denominator of the fraction.
4. How many miles can the automobile travel on a full tank of gasoline? Show your work by writing the **quantity** from CTQ 2 multiplied by the appropriate **fraction** from CTQ 3. Show all units.
5. Explain why the **answer** to CTQ 4 does not include the unit "gallons."
6. Explain why the **fraction** used in CTQ 4 may be called a **conversion factor**.
7. Do all four conversion factors below give equivalent information? Explain your answer.

$$\frac{27\text{ mi}}{1\text{ gal}} \qquad \frac{27\text{ mi}}{\text{gal}} \qquad \frac{1\text{ gal}}{27\text{ mi}} \qquad \frac{1/27\text{ gal}}{\text{mi}}$$

Model 2: Definitions of the inch and the foot

- 1 inch = 2.54 cm (exactly)
- There are exactly 12 inches in one foot.

Figure 1. Excerpt from ChemActivity 4, Unit Conversions: Metric System. (Adapted with permission from reference 16. Copyright 2007 John Wiley & Sons.)

Critical Thinking Questions:

8. How many centimeters are in one inch?
9. Draw a large X through each **incorrect** conversion factor below.

$$\frac{1\ \text{cm}}{2.54\ \text{in}} \qquad \frac{2.54\ \text{cm}}{1\ \text{in}} \qquad \frac{2.54\ \text{in}}{1\ \text{cm}} \qquad \frac{1\ \text{in}}{2.54\ \text{cm}}$$

10. Suppose you want to convert a height from inches into centimeters. Circle the conversion factor in CTQ 9 that you would use. Explain your choice.

Model 3: The unit plan

A unit plan begins with the **unit of the known quantity** and shows how the units will change after multiplying by each conversion factor used, in order. The unit plan for CTQ 10 would be:

in → cm

Each arrow in the unit plan represents one conversion factor.

Critical Thinking Questions:

11. A basketball player is seven feet tall.
    a. Using Models 2 and 3 for reference, complete the unit plan for converting the height of the basketball player into centimeters.

       feet →   →

    b. Write the two conversion factors corresponding to each arrow in part (a).
    c. Perform the calculation by multiplying the **quantity** by each **conversion factor** in order. Show all work with units.

Figure 1. Continued.

Even within this set, compromises must be made regarding depth of coverage. For example, one activity has the grand title of "Reactions of Organic Molecules," perhaps implying that everything about organic reactions can be contained in an activity requiring 45 minutes of class time. Of course, this is not the case, as the activity in fact contains only a brief introduction to seven types of reactions that have particular importance in biology. If this activity were to be used in one course of a two-semester sequence, it could be used as the introductory activity for organic reactions, to be followed by more activities exploring the details of these or additional reactions.

An additional compromise made due to time constraints in the one-semester GOB course is the inclusion of a greater number of concept formation style activities rather than concept invention style activities. Concept invention activities are those in which students inductively develop their own idea about a concept before a name is attached to it; concept formation activities explicitly present and then explore a concept (*21*). While concept invention activities most closely follow the learning cycle, concept formation activities are easier to write and require less classroom time to complete. That is, the explicit presentation and subsequent exploration of a concept requires less class time, because students are not required to analyze data and develop their own ideas about a concept before it is defined. Unfortunately, the reduction of the fraction of concept invention activities somewhat short-circuits the development of some of the process skills that the POGIL model is designed to develop. In a two-semester sequence, more class time (in the form of either additional or alternative activities) can be devoted to each topic.

Classroom Practice

The GOB course is normally the sole physical science course taken by health science majors. Most students surveyed indicate that they are taking the course only because it is required in order to apply to their program of interest (nursing, dental hygiene, *etc.*) and most enter the course with no particular interest in the subject matter. Anything that might be thought to hinder their goal (passing the course with a grade of "C" or better) is met with disdain, and this may include innovative teaching methods.

The students need to be convinced, then, both that the knowledge and the skills they will gain in chemistry will be useful to them, and that the POGIL method can increase their chances of gaining the knowledge and skills that they will need. Providing them with specific data and examples of student success in introductory science courses employing methods other than the traditional lecture (*1-5, 18*), and stressing the importance of analytical and communication skills in future courses in their field and in their careers, can help overcome initial student hesitancy.

As might be expected, careful selection of the three to four students in the learning teams is particularly important. (Four is the preferred number, allowing for a team of three if one member is absent.) Selecting students with a diversity of backgrounds, especially with regard to previous success in science and mathematics courses, is advisable. Teams may be varied for the first few days or weeks to give students a chance to discover who they work well with; this author has found that after this period, keeping teams intact has worked well. These guidelines follow the recommendations of Johnson, Johnson, and Smith (*22*), in which they make the case that the relationships built within a permanent "base group" are important. Indeed, even teams that initially seem to be somewhat dysfunctional usually develop the positive interdependence and support needed for success in this method, and when surveyed, express their preference for remaining in the same teams in which they have learned to be successful.

One classroom approach that is sometimes used by instructors is to integrate a small number of POGIL activities into a lecture course, and this has been done successfully. But if a significant portion of the content is delivered both with this sort of didactic approach and also with POGIL activities (in either order), students may begin to undervalue the guided-inquiry activities. For this reason, an instructor may choose to use a classroom activity even when no new content is to be developed that day. In this way, students who are beginning to buy into the legitimacy of the guided inquiry method can continue using class time in the way in which they have become familiar—for example, a 5-minute quiz or introduction, 40 minutes working on the activity, and a 5-minute wrap-up—even when no new concepts are introduced. Examples of this type of activity include practicing solving stoichiometry problems or learning to recognize a list of organic functional groups.

The one-semester GOB course in particular is very fast-paced, with each class session often exploring a different content-heavy topic. As the course proceeds, instructors should take advantage of every possible opportunity to remind students of how far they have come, what they have learned, and how it applies to their future—since without guidance, students tend to see each activity or topic as a separate item and miss the "big picture." For example, before major examinations, students often appreciate summary or review sessions, which help them to gauge the progress they have made, and to assess whether they have been able to identify and master the important concepts in the activities.

Student Outcomes

Since there are limited numbers of GOB courses utilizing the POGIL method, the data indicating success are also limited—but the existing evidence is encouraging. When moving from a lecture-based to a POGIL instructional method, student performance in the GOB course (*23*) and on standardized exams

(*24*) has remained constant or increased slightly. Students surveyed are just as unhappy with the pace of the one-semester course (always the main complaint in this class), but are happier with the immediate feedback they receive. Some students have expressed gratitude for teaching them a method that they can apply even in their other courses which do not have the daily feedback, but in which the students are expected to digest large chunks of material and have only a few midterm exams on which to display their mastery of it (*24*).

Future Directions

The sequence of activities shown in Table I follows the general outline of most GOB textbooks, though many textbook topics are necessarily omitted or condensed. In a two-semester sequence, more time may be taken to explore each activity. Additionally, the activities could be expanded or used as introductions to topics and supplemented with additional activities on the same or other topics. However, in the attempt to make the course more relevant to the clientele, a more integrated approach seems useful. Integration of traditional organic chemistry topics into other sections of the course would not reduce the content, but would reallocate it in a more appropriate and palatable way. Early attempts at such integration have met with some success (*13-14*). Such integration meets with the recommendations of the ACS task force reported in 1984, "To depart from the traditional compartmentalization of inorganic, organic, and biological chemistry and emphasize integration of basic concepts throughout the course" (*11*). Unfortunately, in the two decades since the publication of this report, standard textbooks are still traditionally compartmentalized.

If some consensus among the educational community were to develop regarding the content and organization of the GOB course, some topics could be omitted or condensed, and others could be investigated in greater depth. Ideally, a textbook would be available to coordinate this integrated approach with a set of POGIL activities designed to accompany that text.

Care must be taken, however, not to eliminate topics that might initially seem to be unnecessary in the GOB course. In interviewing practicing nursing instructors, one topic that seems not to be mentioned by the nurses as being essential is the concept of stoichiometric calculations in chemical reactions. However, the same instructors recognize that success in chemistry is a good predictor of success in certain nursing courses, such as pharmacology. Discussions with students seem to indicate that while the concept of stoichiometry itself may not be directly applicable to the study or practice of nursing, the *skills* developed by solving this type of problem are indeed quite useful (for example, when performing dosage calculations in a pharmacology course).

Conclusions

The use of POGIL in the GOB course presents unique challenges, but these challenges can be met. POGIL activities that address both the specific needs of students in this course and of a syllabus with a comprehensive topic list can and have been prepared. Opportunities still exist for the expansion of these activities, preparing activities that integrate related topics in the course and activities on additional topics, instructing more facilitators in the use of POGIL, and for more comprehensive assessment of the effectiveness of POGIL in the GOB and other nonmajors courses. If the community of chemistry educators continues to move towards a more integrated approach to the topics in the GOB course sequence, the POGIL community can and should participate in and coordinate with that reform.

References

1. Farrell, J. J.; Moog, R. S.; Spencer, J. N. *J. Chem. Educ.* **1999**, *76*, 570-574.
2. Hanson, D. M.; Wolfskill, T. *J. Chem. Educ.* **2000**, *77*, 120-130.
3. Hinde, R. J.; Kovac, J. *J. Chem. Educ.* **2001**, *78*, 93-99.
4. Lewis, J. E.; Lewis, S. E. *J. Chem. Educ.* **2005**, *82*, 135-139.
5. Straumanis, A.; Bressette, A. Simons, E. *unpublished results.*
6. Moog, R. S.; Farrell J. J. *Chemistry: A Guided Inquiry*, 3rd ed.; John Wiley & Sons: New York, 2006.
7. Hanson, D. M. *Foundations of Chemistry: Applying POGIL Principles*, 2nd ed.; Pacific Crest: Lisle, IL, 2006.
8. Moog, R. S.; Spencer, J. N.; Farrell J. J. *Physical Chemistry: A Guided Inquiry Atoms, Molecules, and Spectroscopy*, 1st ed.; Houghton Mifflin: Boston, MA, 2004.
9. Moog, R. S.; Spencer, J. N.; Farrell J. J. *Physical Chemistry: A Guided Inquiry Thermodynamics*, 1st ed.; Houghton Mifflin: Boston, MA, 2004.
10. Straumanis, A. R. *Organic Chemistry: A Guided Inquiry,* 1st ed.; Houghton Mifflin: Boston, MA, 2004.
11. Treblow, M; Daly, J. M.; Sarquis, J. L. *J. Chem. Educ.* **1984**, *61*, 620-621.
12. Daly, J. M.; Sarquis, J. L. *J. Chem. Educ.* **1987**, *64*, 699-702.
13. Frost, L; Deal, S. T. *Making the Most of a One-Semester GOB Course.* Presented at 18th Biennial Conference on Chemical Education, Iowa State University, July 21, 2004, paper S566.
14. Suits, J. P. *GMOB: A metabolic framework for teaching organic chemistry and biochemistry to health science majors.* Presented at 229th ACS National Meeting, San Diego, CA, March 16, 2005, paper CHED 1379.

15. American Dental Association, http://www.ada.org/prof/ed/accred/standards/dh.pdf (Last accessed on Dec. 18, 2006).
16. Garoutte, M. P. *General, Organic, and Biological Chemistry: A Guided Inquiry*; John Wiley & Sons: New York, 2007.
17. Abraham, M. R. *Inquiry and the Learning Cycle Approach.* In *Chemists' Guide to Effective Teaching*; Pienta, N. J.; Cooper, M. M.; Greenbowe, T. J., Eds.; Pearson Prentice Hall: Upper Saddle River, NJ, 2005; pp 41-52.
18. Hake, R. R. *Am. J. Phys.* **1998**, *66*, 64-74.
19. Walhout, J. S.; Heinschel, J. *J. Chem. Educ.* **1992**, *69*, 483-487.
20. Dolter, K. *Chemistry in Nursing Curricula.* Presented at 18th Biennial Conference on Chemical Education, Iowa State University, July 21, 2004, paper S433.
21. Hanson, D. M.; *Instructor's Guide to Process-Oriented Guided-Inquiry Learning*; Pacific Crest: Lisle, IL, 2006; pp 5-6.
22. Johnson, D. W.; Johnson, R. T.; Smith, K. A. *Active Learning: Cooperation in the College Classroom*; Interaction Book Company: Edina, MN, 1991.
23. Eaton, L. *The Effect of Process Oriented Guided Inquiry Learning on Student Achievement in a One Semester General, Organic, and Biochemistry Course.* Presented at 19th Biennial Conference on Chemical Education, Purdue University, August 2, 2006, paper P569.
24. Garoutte, M. P. *Development and implementation of guided-inquiry activities for allied health chemistry.* Presented at 229th ACS National Meeting, San Diego, CA, March 13, 2005, paper CHED 52.

Chapter 12

POGIL in Chemistry Courses at a Large Urban University: A Case Study

Suzanne M. Ruder and Sally S. Hunnicutt

Department of Chemistry, Virginia Commonwealth University, 1001 West Main Street, Richmond, VA 23284–2006

POGIL was implemented in general and organic chemistry courses at Virginia Commonwealth University, a large public, urban research university. These courses ranged in size from 100-250 students and were held in fixed-seating lecture halls filled to capacity. A number of strategies were developed and adopted in order to utilize POGIL in these classrooms. The strategies and the results, including student outcomes and evaluations, will be outlined in this paper.

Virginia Commonwealth University (VCU) is a large, public, urban research university that was founded in 1968. Current enrollment is over 30,000 students, up from approximately 22,000 students ten years ago. Of this population there are over 20,000 undergraduates. The College of Humanities and Sciences, home to the Department of Chemistry, has grown from 6700 to nearly 10,500 students during this time. VCU is the largest employer in the city of Richmond and the ninth largest in the state of Virginia.

The student body at VCU is diverse in all aspects including age, race, ethnicity, background, ability, and part-time status. The average SAT score for VCU students is 1091, with a 25th/75th percentile range of 230 points. The student population is 88% in-state, 82% commuter, 30% part-time, 61% female, 35% minority, and 48% non-traditional college age. Due to the recent, rapid growth and limited space, courses are filled to capacity, and students cannot register for all required courses each semester. This has led to low graduation rates: VCU has a six-year graduation rate of 42%. D-F-Withdrawal rates are fairly high for large classes, including introductory and general chemistry (25-44%), introductory and general biology (32-60%) and college algebra and pre-calculus (30-52%).

The high D-F-Withdrawal rates in many large introductory courses caused us to consider whether lecture was effective at helping to retain students. In addition, large classes made it increasingly difficult for the authors to connect with students as individuals. Moreover, our industrial contacts from Wyeth Pharmaceuticals stressed the importance for their employees to be able to work in teams and to solve problems. The authors expected that adopting the POGIL pedagogy (*1*) would tackle each of these concerns, thus helping students learn more, learn better, and learn more persistently.

Large Lecture Class Characteristics and Their Effects on POGIL Implementation

Many characteristics of large classes make POGIL implementation seem prohibitively difficult. These characteristics are discussed at length in Chapter 6 of this volume. The challenges can be addressed and met, as shown in that chapter, and as we will demonstrate herein. This section highlights some of the more important challenges.

Typically the large lecture classroom is made up of fixed seating in tiered lecture halls with long, closely-spaced rows. Classrooms are generally filled to capacity, making communication between groups and the facilitators difficult at best. With limited group-facilitator interaction, the facilitator cannot monitor whether students work in groups and follow roles. It is impossible for students to report out answers on the board, and only a fraction of groups can orally report answers. Since the student body is diverse, and there is a high student:faculty ratio (200:1 or higher), students feel anonymous. This often leads to problems with attendance (arriving late or leaving early because no one is watching) and classroom civility (goofing off, inactivity, chatter etc.). Handing out and collecting papers is time-consuming; reviewing reports after class is so time-consuming that it is next to impossible to accomplish.

General Chemistry

The general chemistry course structure at VCU is likely similar to that at other large research universities. There are multiple lecture sections, each taught by different lecturers; the course meets for four hours per week including three hours of lecture and one hour of recitation (breakout sections); recitations are taught by graduate teaching assistants (GTAs); lecture sizes range from 90-300 students; and recitations range from 30-50 students.

Traditional recitations involve problem solving by a GTA, followed by a 20-30 minute quiz. In a traditional lecture, the instructor delivers course material in the form of lecture notes. Some instructors use PowerPoint; some use personal responders (clickers); most ask students to work problems during class; and all use Blackboard (online course management software), primarily to communicate with students or keep track of grades. All instructors used the same textbook (McMurry and Fay).

In the fall of 2003, one General Chemistry I instructor with eight large recitation sections agreed to use POGIL during recitations. Two facilitators were used in each recitation section. This implementation of POGIL is similar to that at Stony Brook University (*2*) and the University of South Florida (*3*). In the spring of 2004, three instructors continued using POGIL during recitations for General Chemistry II (two lecture sections) and Introductory Chemistry (one lecture section). That same semester one instructor used POGIL for recitations and during regular class, reducing lecture to 40% of class time. Enrollment in this class was 60, with two recitation sections.

From fall 2004 to the present, approximately one fourth to one half of students in Introductory Chemistry (i.e., preparatory chemistry, required for students who do not place into General Chemistry) and General Chemistry I and II were taught by instructors using POGIL during recitations. Undergraduate and graduate student facilitators worked with these students during recitations. A new one-credit special topics course was offered to train undergraduate students to work with graduate students as facilitators during recitations. These facilitators met weekly outside of class for training and discussion in the special topics course. This special topics course was approved in fall 2005 as a one-credit, upper level elective titled *Guided Inquiry in Chemistry*. The course is offered both fall and spring semesters.

In the fall of 2005, one instructor used POGIL both in recitations and during class time. Enrollment in this class was 206 at the beginning of the semester, with five recitation sections. Lecture was reduced to about one-third of class time. New concepts and topics were introduced using POGIL activities that students worked on in collaborative groups.

Adoption Strategies for the Large Classroom

A number of strategies were adopted in order to use POGIL in the large general chemistry class. These strategies derive from those described in Chapter 6, *POGIL Implementation in Large Classes.* All the lecture activities were short: a single model, four to six questions, ten to twenty minutes long, and ending with two to four exercises to do at home or with time remaining. The length of the activities was well-suited to the comparatively short class time (50 minutes). Longer activities were given to students in recitations; these were two to four pages long and often included several models. All activities were written by the instructor, but the activities, especially the longer ones, were modeled after the activities in Moog and Farrell (*4*). At the beginning of the semester, the instructor gave a brief rationale for using collaborative, guided inquiry learning. and repeated this as needed when students would ask for more lecture. The instructor stated that collaboration on guided inquiry activities makes the class 'feel' smaller; provides practice at skills employers are looking for; and gives students the experience of learning chemistry in the same way chemists do scientific research.

The instructor gave pre- and post-activity mini-lectures using a tablet-pc, overhead projector or PowerPoint presentations. Lectures prior to the activity often addressed student questions from reflections turned in upon completion of a previous activity. Also, the pre-activity lectures provided the opportunity for problem solving. The PowerPoint presentation included multiple personal responder ('clicker') questions (*5*), which quizzed students on concepts from the previous activity. For example, students completed an activity that introduced solutions, dilution, and molarity during recitation. In the subsequent lecture class, they were given the clicker question shown in Figure 1.

The clicker questions also introduced concepts included in upcoming activities, often soliciting misconceptions. For example, students were asked to predict which had the larger atomic radius, sodium or chlorine atoms, prior to a lecture activity on shielding. An overwhelming majority of students selected chlorine as the larger atom; after the activity, most changed their minds and selected sodium.

Post-activity lectures were tied to the just-completed activities. Three or four groups were asked to write out their answers to selected activity questions on overhead transparencies. These answers were used as 'clicker' questions (eg., which answer is best?) following class-wide discussion. Clicker questions that had elicited misconceptions (i.e., sodium versus chlorine size) were asked again. Applications or skill development exercises, such as those found at the end of an activity, were presented using transparencies or the tablet-pc.

Classroom demonstrations were often presented as part of the activity model. For example, an electric light was placed in a liquid (water, methanol, and glacial acetic acid) or an aqueous solution (aqueous sodium chloride, acetic acid, methanol and sugar). The light turned on if ions were present. The activity started with a table listing these liquids and solutions. Through concept questions, students were led to develop the ideas of strong, weak, and non-electrolyte solutions. A second model in the same activity used molecular cartoons (see Figure 2) to link the students' observations and the new vocabulary to a molecular view of electrolytes as well as strong and weak acids. Finally, a

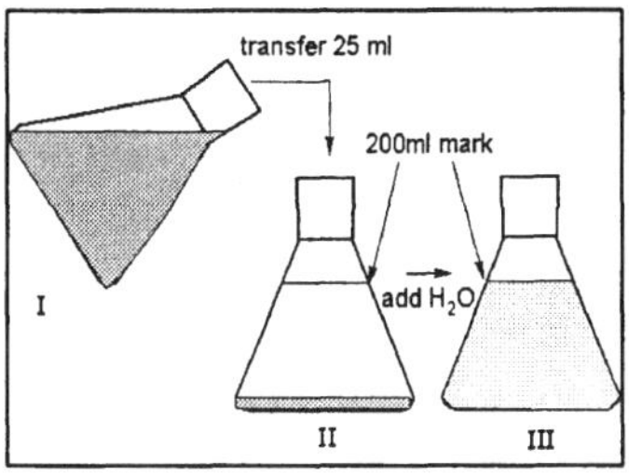

Which two flasks have the same number of moles of solute?
A. I & II
B. I & III
C. II & III

Figure 1. Clicker question from General Chemistry class

subsequent activity developed the concept of acid-base chemistry and the associated skill of writing acid-base reactions.

In another in-class activity students flipped coins as an introduction to simple mass spectrometry. Using clickers, students indicated their results for two coin flips (two heads, two tails, or one of each). Following a few questions related to probability, students were shown mass spectra of Br_2 and then Cl_2. Ultimately they interpreted the mass spectra of halogen compounds like PCl_3.

Students were assessed frequently using 'clicker' questions and on-line homework, rather than paper quizzes. On-line homework was due about every other week. The LON-CAPA, Learning Online Network, Computer Assisted Personalized Approach was used for online homework. LON-CAPA allowed each student to have a unique problem set. Students had multiple attempts at answering questions. The types of questions ranged from numerical answer, string response (naming), predicting reaction products, to graphing and standard matching or multiple choice.

Student Facilitator Training

Facilitator training was critical for undergraduate and graduate students assisting in recitation and lecture. All undergraduate facilitators enrolled in the new course (mentioned above), Guided Inquiry in Chemistry (CHEM350, 1 credit), and all Graduate Teaching Assistants were required to attend. In the first several classes the students participated in a mini-POGIL workshop. They were introduced to inquiry-based, collaborative learning, and group roles and critical processing skills were discussed. The facilitators then broke into teams of three to four students and worked on the upcoming activity, with the instructor modeling facilitation. The undergraduate and graduate student facilitators continued to meet weekly to both go over activities and discuss 'facilitation' such as handling questions from teams, dealing with late students, promoting good teamwork within groups, and helping students clear misconceptions. The facilitators were required to read and discuss two chapters from *How People*

Learn: Brain, Mind, Experience, and School (*6*). In addition, they wrote weekly journal entries based on their experiences with facilitation. In a given week, facilitators were asked to focus on one or two specific issues, such as a particular critical process skill.

Indicators of Effectiveness

Implementation of POGIL in large general chemistry courses has been successful, as measured by exam scores and by student attitudes (discussed in the conclusion). Figure 3 shows exam scores for two exams covering three courses. Each course was taught by the author (Hunnicutt) using similar exams (similar problems, different numbers). Two classes had 200 and one had 300 students. Both graphs show the exam score distribution shifts to higher grades for the POGIL class. The effect was greater for the second exam, which was given about halfway through the semester.

Organic Chemistry

The organic chemistry courses at VCU met three hours a week in multiple sections ranging from 100-250 students. Unlike general chemistry, there were no break-out recitation sections for organic chemistry. In sections taught using traditional methods, the instructor delivered notes either on transparencies, white boards, or using a PowerPoint presentation. Some instructors used the personal responder devices (clickers) and all used Blackboard, primarily to communicate with students or keep track of grades. A common textbook was used by all instructors (Wade).

POGIL has been implemented in one section of organic chemistry every semester for three years. Beginning in the fall of 2003 one instructor implemented POGIL for Organic Chemistry I (CHEM301) in a 100 seat lecture. In the spring of 2004 the same instructor used POGIL for Organic Chemistry II (CHEM302) in a 200 seat lecture. This course included many students (>50%) who did not have POGIL as a method of instruction for the previous CHEM301 course. In all subsequent semesters, one of the authors (Ruder) used POGIL in one section of organic chemistry, while two other sections were taught by other instructors using a traditional lecture format. The various sections of organic chemistry have ranged in size from 100-250 students, in fixed seating lecture halls that were filled to capacity. Each semester the POGIL method was fine-tuned in order to address the challenges noted above and previously in Chapter 6. The details listed below represent the best practice for utilizing POGIL in a large organic chemistry classroom.

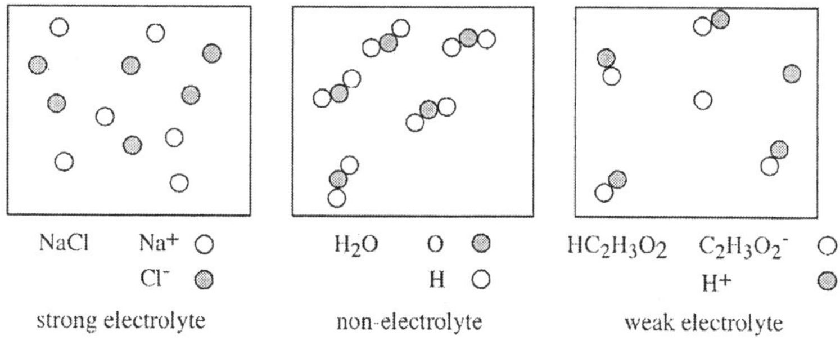

Figure 2. Model 2 from Electrolytes Activity.

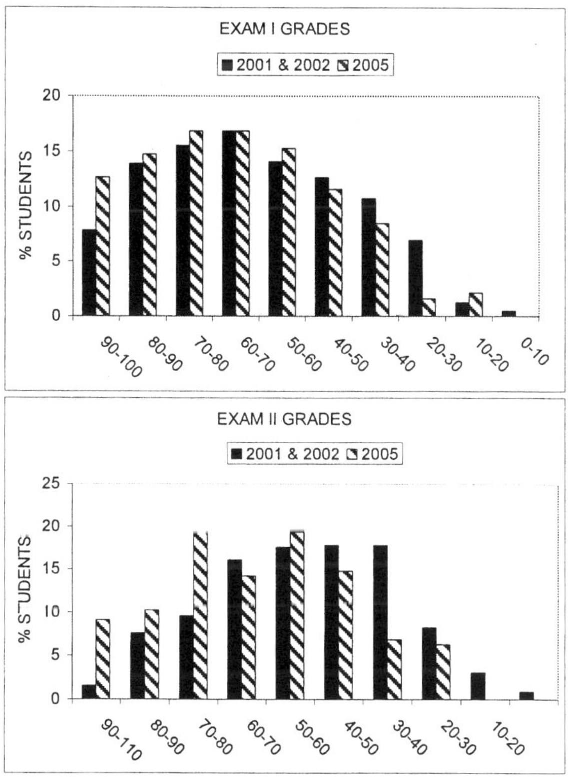

Figure 3. Exams I (top) and II (bottom); comparison of scores given on exams written by author pre-POGIL (2001, 2002, black) and using POGIL (2005, patterned)

Classroom Structure

During drop-add week students worked in groups of three to four on activities during class time. After drop-add week, students were told to commit to a group and asked to write down their group members' names and seat numbers. Based on this information, a seating chart was created and groups were assigned numbers. A folder for each group was then prepared listing the groups' number and names of group members. This method of group management made it much easier to hand out and collect materials. All materials were handed out and returned to group members via their group folder, which took far less time to accomplish than handing out individual items. Group roles were assigned on a daily basis using random methods such as birthdays or birth place. This method also provided a means of having the students get to know a little bit more about their group members.

A typical class was made up of 40-60 groups. In order to help with group management, undergraduate students who had taken the course before were enlisted as teaching assistants (six TAs for a class of 250, and three TAs for a class of 100). These students were not paid but instead were given one independent study course credit. Teaching assistants were chosen if they had received an A or B grade in organic chemistry and were positive about their own POGIL experience. The duties of the teaching assistants were to attend every lecture and facilitate POGIL activities. Student teaching assistants were placed at various locations throughout the room to facilitate learning during group work. All teaching assistants worked through the activities prior to the class session. In addition to attending every class, they were required to hold a help session once a week (two TAs per session) to go over activities or problems. Some teaching assistants made up problems to hand out during help sessions, after receiving approval from the instructor. The help sessions were instrumental in gaining student trust of the teaching assistants. The undergraduate teaching assistants were not allowed to grade or to have access to the online gradebook. They did help hand out materials, organize tests and quizzes for the folders and score IFAT quizzes (see below) for each group.

Classroom polling devices (*5*) (clickers) were also used to manage the groups. Clicker questions were posed at key intervals during each class activity. The questions were closely related to items from the activity, so students needed to complete a section of the activity in order to answer the question. In this manner groups were forced to catch up if they had fallen behind. After the question was posed, the responses of the class appeared as a bar graph. The correct answers to the questions were not automatically indicated in order to open class discussions on a topic for which a large number of students chose the incorrect answer. The polling devices allowed the instructor and her students to monitor student understanding of a concept.

Quizzes were given after each class activity was completed, approximately once a week. To enhance group-work, group quizzes were given about half of the time during the semester. Each group received one printout of the quiz and one answer sheet. The manager read the quiz question, the recorder recorded the answer, and the other group members made sure that each member had a say in choosing the answer. Group quizzes were given using IFAT multiple-choice testing forms available from Epstein education (*7*). Students scratched off their selected answers, as on a lottery ticket. Since the correct answer was labeled with a star, students immediately saw whether their answer was correct. If not they continued to scratch off answers until the correct answer was revealed. Using this technique, groups received partial credit for getting the correct answer on the second attempt and full credit for getting the answer on the first attempt. Questions that were contingent on the prior question could be written without penalizing the student for getting the first answer wrong.

The closure document was an important aspect of every group activity. After an activity each group was responsible for turning in the closure document in their group's folder. This procedure improved each group's accountability for their work. Students reported three items they learned from the activity and one question they had about the concepts in the activity. The instructor reviewed these forms prior to the next class; however, the forms were not handed back. After reviewing the forms, the instructor addressed misconceptions and major questions during the next class period. It was often surprising to see what students had questions about; many of these misconceptions had never surfaced when teaching with the traditional lecture format.

Classroom Activities

Class activities, written by the instructor and handed out during class, were concept invention activities; they were generally four pages long with several short models. Published activities by Straumanis (*8*) were used as references. Each model was followed by 3-10 critical thinking questions starting with simple directed questions and then more complex questions. Models were kept short in order to keep the large class on task and to be able to intervene more easily. Activities were written to correspond to key topics critical to the understanding of organic chemistry. Less important topics were covered briefly and assigned as reading and homework in the textbooks' online homework generating system. The last page of every activity included questions to be completed after class or by groups that finished early.

An example of a class activity for CHEM302 was the reactivity of carbonyl compounds as electrophiles, in which four models were introduced. The first model, shown in Figure 4, introduced the basic concept of nucleophilic addition to an aldehyde or ketone. This model was the students' first introduction to the reactions of carbonyl compounds. After the model, students were asked to draw polarity arrows and any resonance structures of the starting carbonyl compound. Based on the resonance structures, they were asked to predict which atom of the carbonyl compound would most likely react with a nucleophile. Next, they were asked to draw curved arrows illustrating the addition of the nucleophile to the aldehyde or ketone. Finally, they were asked to draw several carbonyl compounds, predict the reactivity and give a reason for this reactivity (steric and electronic effects). These five questions made up model 1 of this activity. As the class progressed through model 1, the instructor projected a clicker question that monitored the students' understanding of carbonyl reactivity (see example, Figure 5). Depending on the results of the question (from computer generated histogram) the instructor led a group discussion about the topic and could then either repoll the class or move on to the next model.

Model 2 developed the idea of reacting carbonyl compounds with strong nucleophiles (basic conditions) using the Grignard Reaction as the model (four questions). Model 3 introduced reactions of carbonyls under acidic conditions using weak nucleophiles (three questions). Model 4 presented a table of nucleophiles and products that occur on reaction with carbonyl compounds. A series of questions was asked about predicting acidic versus basic reaction conditions and product formation in general (six questions).

A typical class session consisted of an opening (15 minutes) that included either a quiz and/or an overview of the concepts learned from the previous day's activity and clarification of the students' questions from the closure documents. The next topic (10-15 min) was then introduced using a tablet PC and clicker questions. After the introduction, the students formed into groups to work on an activity. The instructor used clicker questions to help groups progress toward completion and to check student understanding of the material. Based on the results of the questions, the instructor interrupted group work to give a mini-lecture or allowed students to continue with the activity. On some occasions transparencies were handed out to selected groups to write their answers to a specific question. These answers were then discussed with the whole class or the answers were turned into a clicker question. This method motivated students to stay engaged in group work and increased group and individual accountability. The undergraduate facilitators were also available to answer questions and guide groups towards understanding the concepts presented. The facilitators reported back to the instructor on whether groups were having difficulties and how far groups were progressing on a particular activity. In general, class activities took approximately 30-45 minutes, and students were told to complete the activities after class if they did not finish it during the class period. The final five minutes of class was used to wrap up a concept, either with a clicker question or with a mini lecture. Finally, all groups turned in a closure document.

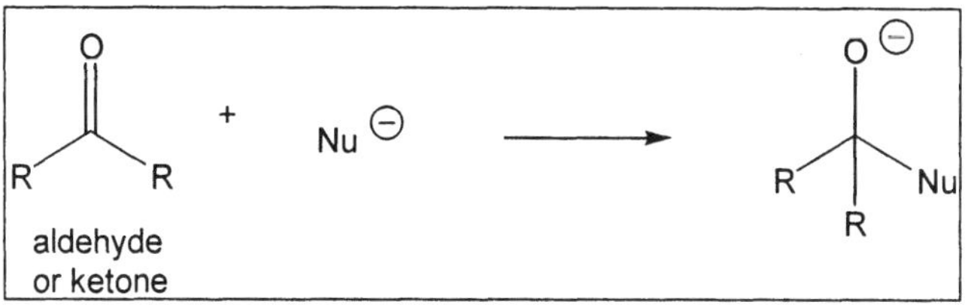

Figure 4. Model on class activity: Addition of nucleophiles to carbonyls

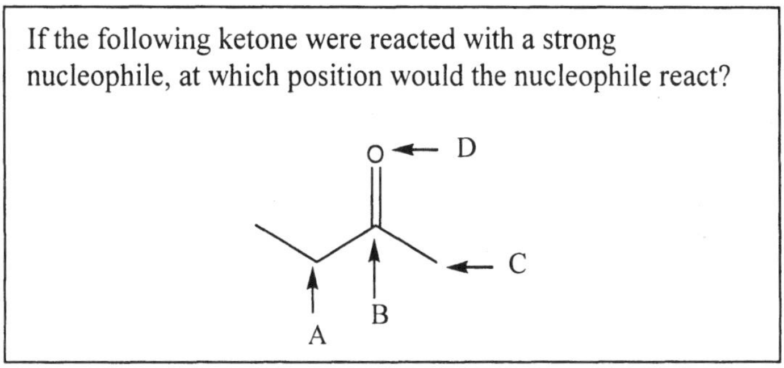

Figure 5. Clicker question for Organic Chemistry class

Indicators of Effectiveness

The following data show that use of the POGIL method in large organic classes was successful. Improvement in test scores as well as retention of material was seen. The following charts in Figure 6 show the results from one instructor (Ruder) over three years of teaching CHEM301. The same instructor taught this course to approximately 100 students in the same room at the same time in the fall semester three consecutive years. The same textbook was used and testing format was identical all three years. The only difference was in the fall of 2002, the instructor taught exclusively using traditional lecture format. The material was delivered in lectures on overhead transparencies. During the fall 2003 and 2004 semesters, the instructor taught predominately using the POGIL method. Clickers were added to help facilitate the activities during the 2004 fall semester. The results shown below are scores obtained on the three exams given during the course of the semester. The black bar illustrates the scores obtained by the students taught in the traditional lecture format. The two patterned bars illustrate the scores obtained by the students taught in the two years of POGIL method. There is little difference in performance on the first

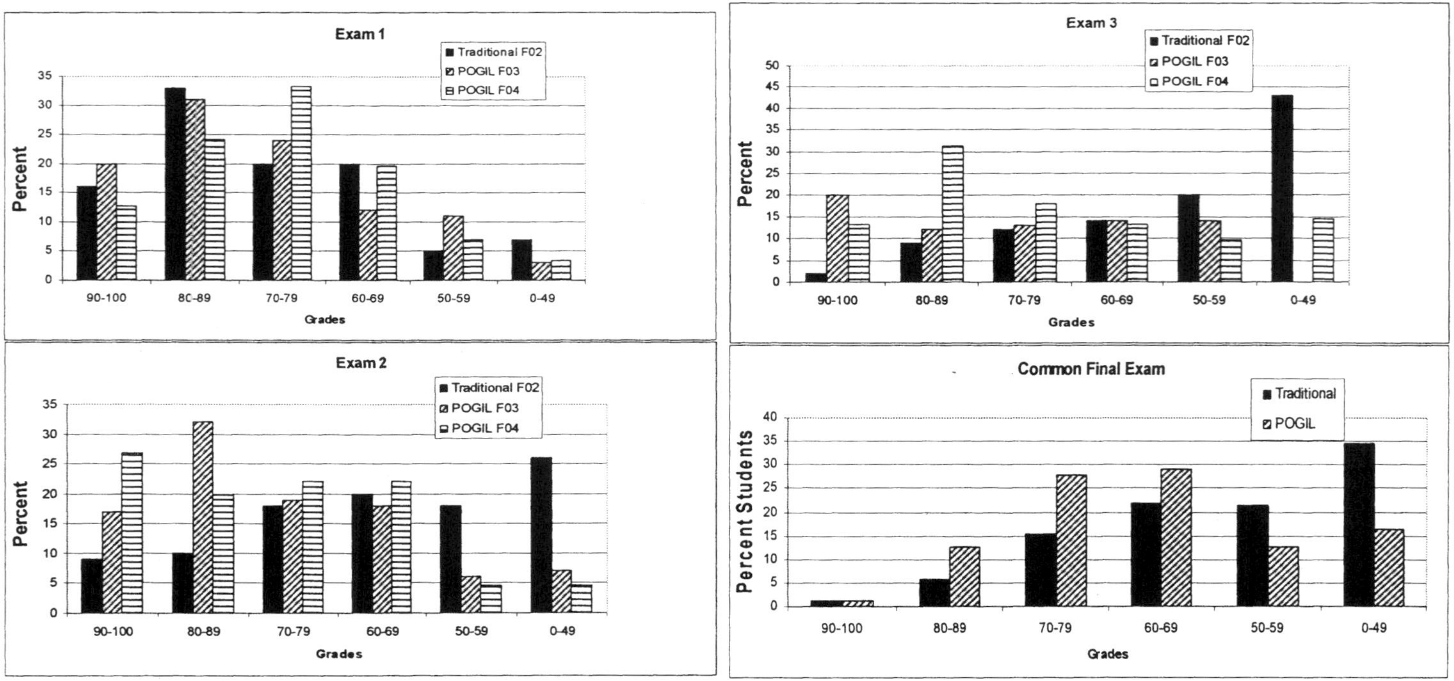

Figure 6. Exam results for CHEM 301. Comparison of scores for exams 1-3 written by author pre-POGIL (2002, black) and using POGIL (2003, 2004, patterned). Comparison of scores for common final exam by other instructors (traditional, black) and author (POGIL, patterned)

exam, which is not surprising since there is a fair amount of review material and since the group work was just in the early stages at the beginning of the semester. However, by the second exam when all material was new and the POGIL method had been in use for almost half of the semester, it can be seen that the POGIL classes outperformed the traditional section. The third exam results were similar; the POGIL sections had more students in the A and B range while the traditional section had more students performing in the D and F range. In general, class averages overall in the POGIL sections were around 10% higher than in the traditional lecture sections. The final chart shows the performance on a common final exam given by two different instructors during the fall 2004 semester. The black bar illustrates the scores obtained on the common exam in the course taught by another instructor using the traditional lecture format. The patterned bar shows scores obtained on the common exam by students in the POGIL section.

The next results (Figure 7) illustrate the amount of retention of organic material at the start of the second semester of organic chemistry (CHEM302). On the first day of class in CHEM302, students were given an assessment quiz on some basic material covered in CHEM301. Students taking the quiz had completed CHEM301 from various instructors. Students having completed CHEM301 in a traditional lecture format with instructors other than the POGIL instructor are shown as a solid bar. Students who took CHEM301 in the POGIL section are shown as a patterned bar. The results of the assessment quiz for two semesters are shown in Figure 7. The only difference in the quizzes is that in the spring 2004 (chart on top) the quiz was written by the author (POGIL section instructor), while in the spring of 2005 the quiz was written by an independent instructor who had not taught any of the students in any of the prior semesters. These results show that the students completing the first semester of organic chemistry in the POGIL class (patterned bar) outperformed all other sections by 25-30%

Conclusions

This case study of large general and organic chemistry courses at VCU demonstrates that, despite numerous challenges, POGIL can be successfully implemented in a large class environment. The affect of the POGIL method of teaching on these courses is dramatic. The authors contend that implementation of POGIL in a large classroom changes that classroom environment more significantly than implementation in a smaller classroom. Our classes were interactive, with students doing work in class. Students were attentive, attendance was good and there was rarely a rush to get out the door at the end of class. As facilitators circled the room, they observed "light-bulb" moments. The POGIL classroom was a good way to obtain feedback and better insight into

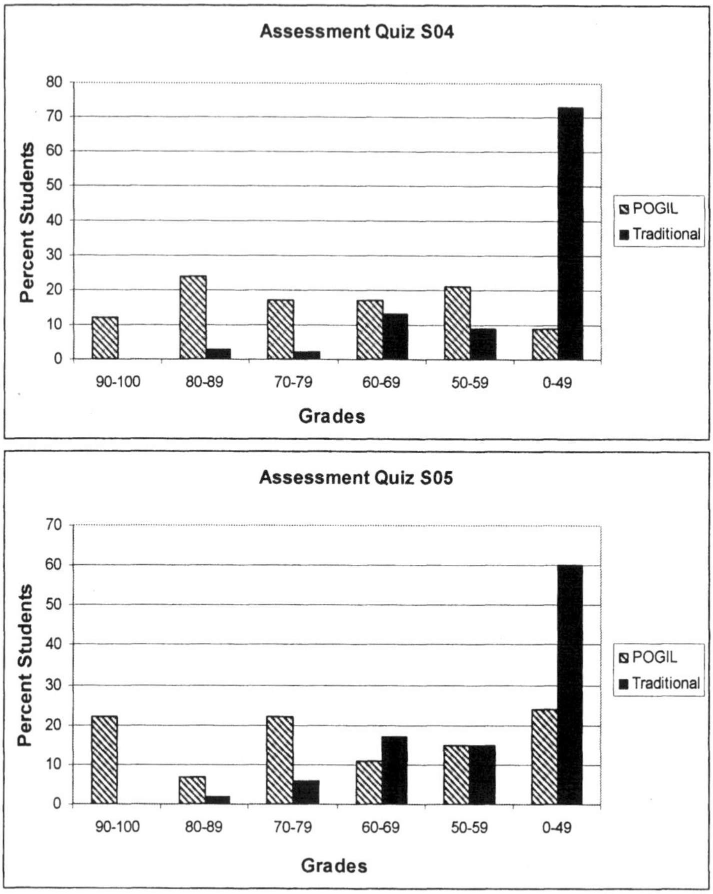

Figure 7. Results on assessment quiz given at the beginning of CHEM 302. Comparison of students having traditional lecture for CHEM 301 (solid) and POGIL for CHEM 301 (patterned)

students' understanding of the material being presented. The instructors interacted with far more students than in a traditional setting. The large POGIL classroom was chaotic at times, but benefits were seen in improved test scores and more A-B-C grades. A majority of students in organic chemistry preferred POGIL over the traditional lecture. They were positive, but they believed they had to work harder in the POGIL classroom. The general chemistry students wanted more traditional lecture, but a majority believed they learned from other students, felt free to ask questions, and clearly understood their responsibilities for the course.

References

1. Farrell, J. J., Moog, R. S., Spencer, J. N. *J. Chem. Educ.* **1999**, *76*, 566-569.
2. Hanson, D., Wolfskill, T. *J. Chem. Educ.* **2000**, *77*, 120-129.
3. Lewis, S. E., Lewis, J. E. *J. Chem. Educ.* **2005**, *82*, 135.
4. Moog, R. S., Farrell, J. J. *Chemistry: A Guided Inquiry*; John Wiley & Sons: New York, NY, 2006.
5. Duncan, D. *Clickers in the Classroom;* Pearson: San Francisco, CA, 2005.
6. *How People Learn: Brain, Mind, Experience, and School;* Bransford, J. D., Brown, A. L., Cocking, R. R., Eds.; National Academy: Wash. DC, 1999.
7. Epstein, M. L.; Epstein, B. B. Brosvic, G. M. *Psych. Reports*, **2001**, *88*, 889-894.
8. Straumanis, A. R. *Organic Chemistry: A Guided Inquiry*; Houghton Mifflin: Boston, MA, 2004.

Chapter 13

POGIL in the Physical Chemistry Classroom

James N. Spencer and Richard S. Moog

Department of Chemistry, Franklin and Marshall College, P.O. Box 3003, Lancaster, PA 17604

Student-centered instruction, a key component of the POGIL pedagogical approach, is not often considered in advanced chemistry courses such as physical chemistry. The reasons for this are, in part, the unsupported belief that advanced courses require direct instruction – that is, a lecture-based approach. Institutions now employing POGIL methods in various advanced courses have demonstrated that POGIL techniques are as applicable to upper level chemistry courses as for general or organic chemistry. Indeed, practitioners of POGIL in physical chemistry generally agree that this course readily lends itself to an active classroom environment. This chapter provides an example of a POGIL classroom activity for a typical junior-level physical chemistry course, and offers some details on classroom implementation.

Recent research (*1*) has shown that an effective learning environment is one in which the students are actively engaged, an environment where there is something for students to do. In a student-centered classroom, the focus is on the learning of the students rather than the instruction of the teacher. The instructor acts more as a facilitator of learning, asking probing questions to help guide the students to develop understanding and address misconceptions or misunderstanding. Many instructors, particularly those teaching advanced courses such as physical chemistry, find it difficult to see how the sophisticated and difficult content of their course can be mastered using a student-centered approach, particularly given the time constraints under which they must operate. In this chapter, we show how the POGIL approach can be successfully implemented in a junior level physical chemistry course.

Principles of POGIL

A brief review of Process Oriented Guided Inquiry Learning (POGIL) is given here; additional details may be found in the first chapter of this volume. Central to this pedagogy is the perspective that each student constructs his or her own knowledge, and that this construction depends upon what the student already knows. In addition to this mastery of content and understanding, emphasis is also placed on the development of important learning and other process skills, such as communication, critical thinking and problem solving. In the POGIL paradigm, instructors facilitate learning rather than serving as a source of information, and students work in small self-managed groups on materials specially designed for this approach. In general, these activities guide students to develop the important concepts of the course by using a learning cycle structure. The learning cycle (*2*) is an inquiry-oriented instructional strategy that consists of three general steps. First there is an exploration involving data, a model, or other information from which the student is guided to the second step, the construction or formation of the concept intended. The third step is an application of what has been learned. Thus, the premises of the POGIL philosophy are that students will learn better when they are actively engaged and thinking in class. They construct knowledge and draw conclusions themselves by analyzing data and discussing ideas. They learn how to work together to understand concepts and solve problems. Additional information is available from the POGIL website (*3*).

Implementing POGIL in the Physical Chemistry Classroom

One of the most important aspects of a successful POGIL implementation is the careful melding and interplay of the learning environment with the specially designed POGIL activity. Although this will necessarily vary from institution to institution and instructor to instructor, depending on the size of the class, the physical environment of the classroom, the background of the students, etc., many aspects will be similar. We present a typical classroom experience here to better exemplify how this might work. Further details are available elsewhere (*4,5*).

Structured group learning is one of the key aspects of the POGIL learning environment; typically students are placed in groups of three or four, usually with assigned roles and corresponding responsibilities. Often, a physical chemistry activity will begin with a focus question which the groups are asked to answer. In many (if not most) cases, the students do not have a firm chemical basis on which to respond. However, this process forces the students to make a prediction based on whatever they do already know. This accomplishes two things: 1) it gives the instructor a sense of what the students already understand or think they understand – that is, what the students' current construction of the relevant knowledge is; and 2) it forces the students to bring their current understanding out into the open, so that if (when) they confront information or conclusions that are at odds with their current understanding, the "disequilibrium" is made more apparent and tangible. At this point, there is no discussion of what the correct answer is; rather, the various group perspectives can be shared, perhaps with brief explanations, and then the class proceeds, with the understanding that completing the activity will bring a resolution to the focus question. At this point, the instructor may provide a brief (2-3 minute) discussion designed to place the day's activity into some context, and then the groups begin work. The instructor moves among the groups listening to their discussions, and intervening sparingly. Occasionally, particularly if common difficulties are observed, additional brief (less than 3 minute) presentations may be made to the entire class. At the end of class, closure is provided by summarizing the main points of the day's activity (either by the instructor or by the students), or with discussion of the original focus question showing how the activity had enabled the students to formulate a more meaningful response to that question.

The Structure of an Activity

An example of a typical thermodynamics classroom activity is given at the end of this chapter. Here we give a somewhat detailed description of how the activity is structured. This activity is encountered more than halfway through a semester of a typical junior level thermodynamics course. Students have previously encountered similar activities on real and ideal gases, the first, second and third laws, Gibbs energy, phase equilibrium, and an introduction to ideal solutions. In these prior activities, the students are guided to develop the important concepts, with the accepted terms describing these ideas often being introduced *after* the concept has been developed. In the example here, the chemical potential for a component of an ideal solution is developed.

The activity begins with the previously mentioned focus question. The students are expected to answer based on their current understanding; they will actually develop an answer to this question as they work through the activity. The focus question is one that the students have not discussed previously, nor are they expected to have read about it in the text ahead of time. Thus, they have some difficulty in quickly generating a "correct" answer with any certainty – within the 30 to 60 seconds allotted for them to discuss the question.

Model 1 reminds them of what they have established previously, introduces the term *vapor phase* as standard terminology, and provides a representation (Figure 1) of a pure liquid in equilibrium with its vapor. The Information section provides additional standard terminology for concepts, many of which have been developed previously. This presentation of information begins an "Exploration" phase of this activity, continuing with the first few Critical Thinking Questions (CTQs) which direct the students back to the Model, Figure, and Information. CTQ 1 is designed to develop information processing skills, insuring that the students have examined the Model, Figure, and Information and understand what they mean, including the relationship between the chemical potential of the pure liquid at 298 K and its vapor, $\mu_{A(l)}{}^{*} = \mu_{A(vap)}$.

The students have previously developed the relationship for gases $\mu_{A(vap)} = \mu_A^{*} + RT \ln P_A$. In CTQ 2, the students use this relationship, but must also recognize that the pressure is that of the pure liquid, $P_{A(vap)}{}^{*}$, so that $\mu_{A(vap)} = \mu_A^{*} + RT \ln P_A^{*}$. Thus, newly introduced information is being combined with previous knowledge to reinforce and also extend student understanding. The "Exploration" continues with Model 2, in which an ideal mixture is introduced. CTQ 3 requires students to use the information supplied in Model 1 to recognize that the Gibbs energy relationship must also apply to mixtures. CTQ 4 guides students to generalize and conceptualize how their relationship developed in CTQ 2 can be applied to mixtures. Then, in CTQ 5, Raoult's Law (developed in a previous activity) is used to obtain an expression that includes a composition variable. An explanation such as that requested in CTQ 6 requires students to put into their own words their understanding in clear and concise language, an important aspect of making sense of what might otherwise simply be a collection of symbols on a piece of paper. This "Concept Invention" phase of the activity culminates with CTQs 7 and 8, in which a general statement of the important relationship is developed and explained.

The Exercises, generally done as homework outside of class, test the students' ability to apply the concepts and ask for an answer to the focus question. This section constitutes the "Application" phase of the learning cycle structure of the activity. If a group finishes early and the instructor decides not to have the group proceed to the next activity, a particularly probing Exercise may be assigned to the group. If several groups do not finish an activity in the allotted time, the instructor makes a decision as to whether the activity could be finished outside the class and, if so, may make this assignment. Sometimes only minimal discussion is needed to bring closure to the activity and the instructor may choose to do this either by calling on groups who have finished or by guiding the students in a whole class discussion to the completion of the activity.

Conclusion

POGIL Physical Chemistry activities are available from Houghton Mifflin for a two-semester course. The Thermodynamics activities (*6*) include Gases, Thermodynamics, Electrochemistry, Kinetics, and Math for Thermodynamics. The Structure and Bonding activities (*7*) include Atomic and Molecular Energies, Electronic Structure of Atoms, Electronic Structure of Molecules, The Distribution of Energy States and Spectroscopy. As of this writing, activities for a full year of general chemistry (*8,9*), a full year of organic chemistry (*10*), and an allied health science (GOB) course (*11*) have been published, along with the preliminary edition of materials for Preparatory Chemistry (*12*); materials for various other courses are currently under development. Up-to-date information concerning available course materials can be obtained from the POGIL web-site (*3*).

References

1. *How People Learn*; Bransford, J. D.; Brown, A. L.; Cocking, R. R., Eds.; National Academy Press: Washington, DC, 1999.
2. Abraham, M. R. In *Chemists' Guide to Effective Teaching;* Pienta, N. J.; Cooper, M. M.; Greenbowe, T. J., Eds.; Prentice Hall: Upper Saddle River, NJ, 2005; pp 41-52.
3. *Process Oriented Guided Inquiry Learning* URL http://www.pogil.org . Last accessed October, 2007.
4. Farrell, J. J.; Moog, R. S.; Spencer, J. N. *J. Chem. Educ.* **1999**, *76*, 570-574.
5. Hanson, D. M. *Instructor's Guide to POGIL*; Pacific Crest: Lisle, IL, 2006.
6. Spencer, J. N.; Moog, R. S.; Farrell, J. J. *Physical Chemistry: A Guided Inquiry. Thermodynamics*; Houghton Mifflin: Boston, MA, 2004.
7. Moog, R. S.; Spencer, J. N.; Farrell, J. J. *Physical Chemistry: A Guided Inquiry. Atoms, Molecules, and Spectroscopy*; Houghton Mifflin: Boston, MA, 2004.
8. Moog, R. S.; Farrell, J. J. *Chemistry: A Guided Inquiry*, 3rd ed.; John Wiley and Sons: Hoboken, NJ, 2006.
9. Hanson, D. M. *Foundations of Chemistry*; Pacific Crest: Lisle, IL, 2006.
10. Straumanis, A. *Organic Chemistry: A Guided Inquiry*; Houghton Mifflin: Boston, MA, 2004.
11. Garoutte, M. *General, Organic and Biological Chemistry: A Guided Inquiry, 1st Edition;* John Wiley & Sons: New York, 2006.
12. March, J.; Caswell, K.; Lewis, J. *Introductory Chemistry Modules: A Guided Inquiry Approach, Preliminary Edition,* Houghton Mifflin Company: Boston, MA, 2008.

ChemActivity **T16**

Chemical Potential for a Component of a Solution

Focus Question: Addition of a solute to a solvent increases the solvent's contribution to the entropy of the liquid phase. Does the solvent's contribution to the Gibbs energy of the liquid phase increase or decrease?

Model 1: Gibbs Energy of a Pure Liquid and its Vapor.

Consider pure liquid A in equilibrium with its vapor. When phases are in equilibrium, if there is a gas phase it is called the vapor(*vap*) phase.

Figure 1: Pure Liquid A at 298 K in Equilibrium with its Vapor

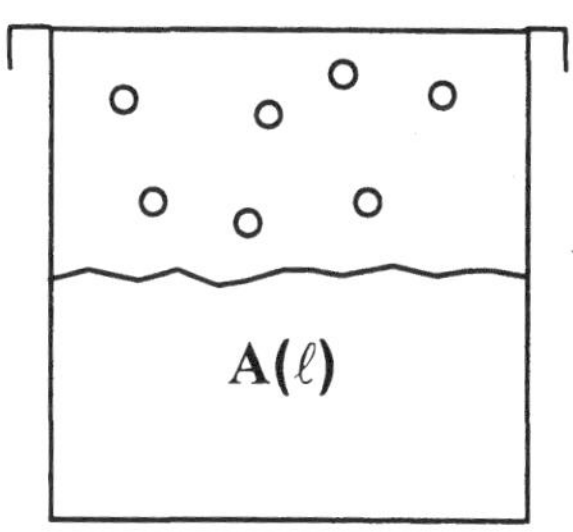

Vapor pressure of pure A = $P^*_{A(vap)}$

When two or more phases are in equilibrium at constant temperature and pressure, the temperatures and pressures of all phases must be the same and the Gibbs energy of each component present in each phase must be identical.

Figure 1. ChemActivity T16. (Reproduced with permission from reference 6. Copyright 2004 by Houghton Mifflin Company.) *Continued on next page.*

Information

The Gibbs energy of pure liquid A is $\mu^*_{A(l)}$.
The Gibbs energy of pure gas A at 1 bar is $\mu^\circ_{A(g)}$.

μ_i is the partial molar Gibbs energy and is frequently called the *chemical potential.* It is a measure at constant T and P of the escaping tendency of a component from a phase. Various symbols are used to represent the chemical potential, μ_i and G_i being the most common.

Critical Thinking Questions

1. For the system at equilibrium described in Figure 1:

   a) what are the two phases present and what is the composition of each phase?

   b) what pressure is exerted on A(*l*)?

   c) what is the temperature of A(g)?

   d) what is the mathematical relationship between $\mu^*_{A(l)}$ and $\mu_{A(vap)}$ at equilibrium?

2. Provide an expression for the $\mu_{A(vap)}$ in terms of $\mu^\circ_{A(g)}$, the temperature T and the vapor pressure $P^*_{A(vap)}$.

Model 2: An Ideal Mixture of Benzene and Toluene.

Consider an ideal mixture of benzene and toluene at equilibrium with its vapor. Assume that the vapor phase behaves ideally also.

Figure 1. Continued.

Figure 2. A Mixture of Benzene and Toluene in Equilibrium with the Vapor Phase at 300 K.

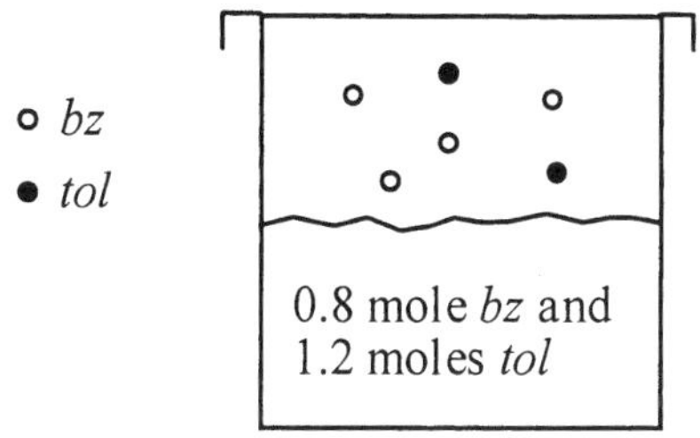

Critical Thinking Questions

3. In Figure 2, what relationship must exist between:

   a) $\mu_{bz(sol)}$ and $\mu_{bz(vap)}$?

   b) $\mu_{tol(sol)}$ and $\mu_{tol(vap)}$?

4. For the solution in Figure 2 provide an expression similar to that in CTQ 2 for $\mu_{bz(vap)}$, the Gibbs energy of benzene in the vapor. Clearly identify all symbols.

5. Use Raoult's Law to show that

$$\mu_{bz(sol)} = [\mu^{o}_{bz}(g) + RT \ln P^{*}_{bz(vap)}] + RT \ln X_{bz(sol)} \qquad (1)$$

6. Identify each term in the bracketed expression in equation (1), and then explain the significance of the entire bracketed term.

7. Show that equation (1) can be simplified to obtain

$$\mu_{bz(sol)} = \mu^{*}_{bz}(l) + RT \ln X_{bz(sol)}. \qquad (2)$$

   Clearly define all symbols.

8. Generalize equation (2) to provide an expression for μ_i, the Gibbs energy of any component i in an ideal solution in terms of its mole fraction, X_i.

Figure 1. Continued. Continued on next page.

Exercises

1. $\mu_{i(sol)} = \mu^*_{i(l)} + RT \ln X_{i(sol)}$

   Describe in words the meaning of this equation.

2. A and B mix to form an ideal solution. What is the Gibbs energy of A in the solution as compared to the Gibbs energy of pure liquid A? How does the Gibbs energy of B(sol) compare to B(pure)?

3. If B is added to pure A what happens to the Gibbs energy of A?

4. Why does the Gibbs energy of the solvent of an ideal solution decrease upon addition of a solute?

5. Assume that the result from CTQ (8) applies to the (clearly non-ideal) solution of salt in water. Use this result to explain (in grammatically correct English sentences) why the boiling point of H_2O is raised when a small amount of salt is dissolved in it.

Figure 1. Continued.

Chapter 14

Enhancing the POGIL Experience with Tablet Personal Computers: Digital Ink in the Learner-Centered Classroom

Christina Mewhinney[1] and Eric J. Zückerman[2]

[1]Division of Science, Eastfield College, Dallas County Community College District, Mesquite, TX 75150
[2]Department of Chemistry and Physics, Augusta State University, Augusta, GA 30904

Digital ink technology was integrated with the active, learner-centered pedagogy, Process Oriented Guided Inquiry Learning. Tablet PC's were used both wired and wirelessly in a classroom to allow more extensive communication. Cooperative learning techniques combined with the tablets made students more active participants in the learning process. Networked tablets allowed drawing of complex chemical structures, and the projection of student work for discussion. Handouts, files, and quizzes were handled electronically. Student opinion surveys suggest this technology increases learning.

Imagine a classroom where students work in small groups huddled around tablet computers, discussing models, answering critical thinking questions to develop their knowledge of chemical concepts, writing answers and drawing complex chemical structures in an electronic workbook on their tablet PC. The instructor monitors each group's progress by acccssing their tablet screen from the instructor's tablet PC, sending messages to individual student groups or writing on their screen as needed to guide their work. One group, who has not been able to reach a consensus on an answer, seeks assistance from another group via online chat, and then carries their tablet to another group across the room for discussion. As the student work progresses, the instructor sees an "interesting" student answer and projects that group's work from their tablet desktop to a screen in front of the room initiating a whole class discussion. Another group makes suggestions and, with the instructor's authorization, modifies the projected answer via their own tablet PC. Students use the tablets to make notes and summarize the key points learned, and upload their electronic workbook to a web server, where the individual group members and the instructor can access it after class. With the click of a mouse handouts are sent to student tablets in electronic form, and students draw complex chemical structures on electronic quizzes returned to the instructor's computer with another mouse click. This describes a typical class in Mewhinney's networked tablet classroom at Eastfield College.

What makes the use of technology in this classroom different? Computers are commonly used in chemistry classrooms now; many instructors use software presentation programs to prepare their lectures and assign homework for students on software tutoring programs. But these common uses are directed toward the individual learner in a teacher-centered classroom. The technology described in this paper is used in an ***active, learner-centered classroom to increase student interaction*** and better manage class time. The classroom described above integrates innovative technology with innovative pedagogy to enhance student learning.

Networking PC's in a learner-centered classroom has been done before. One system, LUCID (*1*), was specifically designed for use with POGIL. The LUCID software contains specific activities for General and Introductory Chemistry for use by groups of students on desktop PC's. However LUCID does not permit drawing chemical structures or diagrams as can be done with tablet technology, nor does it permit student groups to interact in real time with each other or the instructor during class. Also, unlike other programs, the technology described in this paper is content independent, and can be used to enhance communication in any type of student centered classroom pedagogy.

POGIL in a Networked Tablet PC Classroom

A typical POGIL classroom (without the technology described in this article) provides an environment in which the students can "discover" and learn basic chemistry or other disciplinary concepts (*2*). There are many ways to utilize POGIL pedagogy in a classroom but one common model is shown in Figure 1. The class begins with a discussion of the homework assignment, followed by a quiz on previous class's work. The students then work on new content in a guided inquiry activity (often in the form of an activity book), report their results to the class and a whole class discussion ensues. The instructor usually intervenes only to assist the group in arriving at their own answer, but may provide a 3-minute mini-lecture as needed to clarify a major sticking point. The students provide closure to the class by summarizing the key points learned.

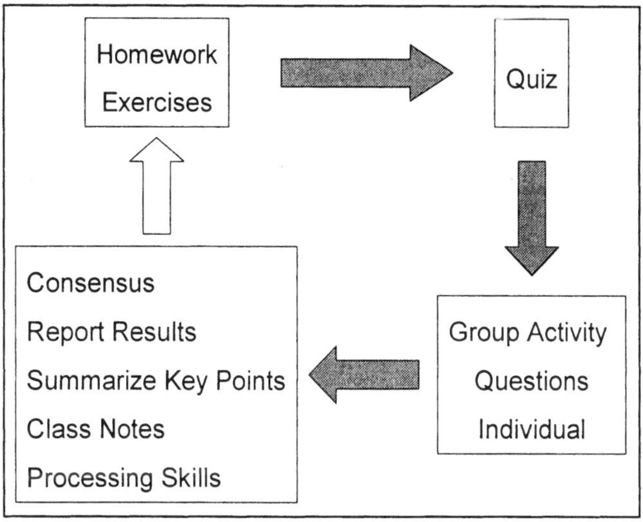

Figure 1. Structure of a typical POGIL class

In addition to teaching chemistry content, the goal of the POGIL pedagogy is to teach processing skills, such as critical thinking, oral presentation, and social and teamwork skills (*3*). Students can learn these skills working in self-managed teams in a cooperative learning environment (*4*). Some instructors assign a formal role with specific duties to each student and rotate these roles among the students in each group. At Eastfield College, General and Organic Chemistry classes are taught by Mewhinney with teams of three persons: a manager, a scribe, and a presenter. After the individual quiz the students work in groups; the scribe records and produces the work product of the group; the presenter speaks for the group in class, explaining the group's thought process and answering questions; and the manager encourages the participation of all, observes and directs the group's work and provides a written summary of their functioning as a group.

The group's main work is to discuss and develop answers to critical thinking questions in a specifically designed learning activity. This is followed by whole class discussion of selected group answers, where "wrong" answers are valued for their learning potential.

All of this can be done without the use of networked Tablet PC's; however, as the rest of this article describes, the use of such technology radically changes the avenues for communication in a POGIL classroom. This is because, from an instructor's point of view, managing a POGIL classroom without additional assistance can be demanding, especially with a larger class, and from an individual student's point of view the challenges are to stay on task, to synthesize and organize class notes and the learning activities, and to function as an integral member of a group.

In pilot classes at Eastfield College during spring and summer 2006 semesters tablet technology was coupled with a wireless network to address areas for improvement in the POGIL classroom suggested by students in Eastfield College class surveys and by faculty in national POGIL workshops (*5*). The areas specifically addressed by the networked tablet project are: the need for the instructor to visit each group to monitor its performance, the minimal communication between student groups, the time required for student groups to report out their results, the numerous times chemical structures are rewritten, the difficulty for the individual student to organize notes from class discussion and workbook activities, and the lack of a final group work product available to the group's individual members and the instructor after class.

The Networked Tablet Classroom

Two recent technological advances are expected to have a significant impact on education in the current decade: these are wireless networking and digital ink. It is obvious that computer networking is a boon to communication and collaboration, and many college campuses have already invested in wireless networks. Many expect the digital ink capabilities used by tablet PC's to open a whole new world for education. This technology has remained relatively unexplored partly because the educational community has been slow to come up with good answers to the question: Why use a tablet when you can type faster?

One exception is organic chemistry. In organic chemistry one does not "type" chemical structures, and it is easier and quicker for most students to draw structures using digital ink than to use chemical drawing software. Structures or parts of structures can easily be copied and pasted, rather than being continually redrawn in each step of a mechanism or synthesis. In General Chemistry students can draw, highlight, modify, and comment on molecular "pictures" or diagrams, as well as complete equilibrium reaction tables, and write out calculation setups. It is for these reasons Mewhinney chose to explore networking tablet PC's in her chemistry classrooms.

With a 2005 Technology for Teaching grant from Hewlett Packard, Eastfield College built a super-smart classroom for chemistry classes (*6*). The hardware in this classroom includes 21 Compaq™ tablet computers, SMART Sympodium™ monitor, wireless network, digital projector and screen. Software used includes Microsoft OneNote™, NetSupport School Pro™, and SMART Notebook™. There are currently alternatives to all of the software and hardware components, and new products are expected to rapidly enter the market.

The classroom was built in stages so as to work out any technical difficulties in increments at each step. The SMART Sympodium™ monitor, Figure 2, was the first piece of equipment installed. This active monitor replaced the monitor on the classroom desktop computer and, coupled with the projection system, allowed the instructor or a student to draw directly on the screen with a special pen. One can draw in any application (including

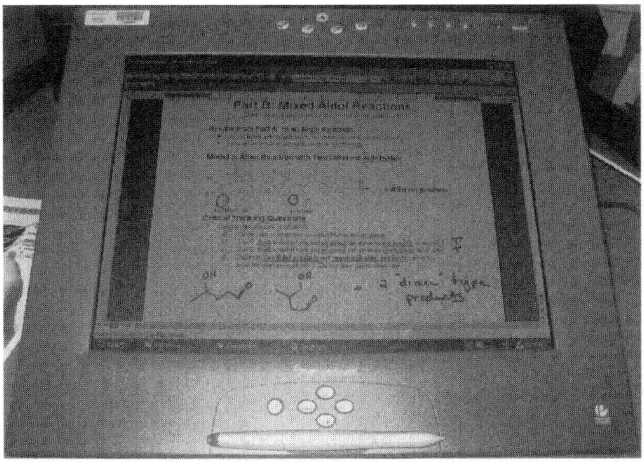

Figure 2. Drawing on a SMART Sympodium™

PowerPoint™) or use the advanced tools provided in the accompanying software, SMART Notebook™. Combined with the projection system the monitor allowed the instructor to draw, project the drawing, and save the drawing as an electronic file.

In the second phase of development one tablet PC was provided to each group of three students during class. The tablets were not initially networked so that any problems with the tablets themselves could be addressed. With permission from the publisher, the graphics in the student activity book, *Organic Chemistry: A Guided Inquiry* (*7*), were printed into Microsoft OneNote™ software. The electronic book was organized into chapters and sections, homework and textbook reading assignments were added, as were spaces to write key concepts learned. The software gives the student the ability to create space anywhere in the document for additional class notes, allows for typing and highlighting as well as drawing, provides many drawing "tools", is word searchable, and can be uploaded to a web server or printed.

In the third phase of the project the group tablets were joined to the wireless network. A subnet was created to reduce the required bandwidth and the response time. NetSupport School Pro™ was installed on all the tablets and the classroom computer. The system was configured and debugged before the remaining tablets were added to the network. This was the most technically challenging phase of the project and tested the patience of the students as well as the instructor. There were many variables in the tablet software configuration that had to be addressed that cause the tablet to drop its network connection.

Once networked, the instructor was able to monitor, project and control all the tablets via NetSupport School Pro™, as in Figure 3. In phase four, the remaining tablets were joined to the network so that each student had access to a tablet for individual electronic quizzes.

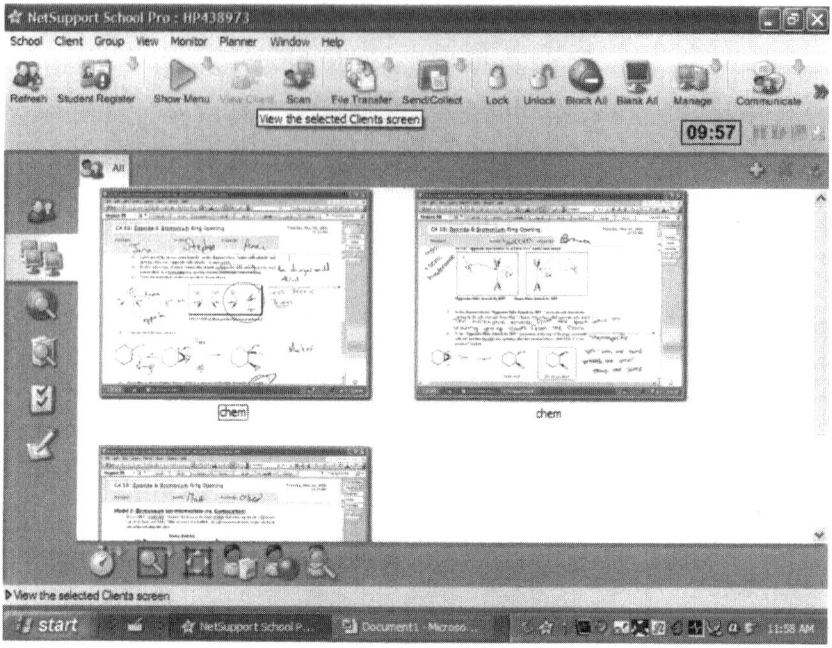

Figure 3. Instructor's desktop monitoring software

The capabilities of this integrated classroom are extensive. Unlike the traditional chalkboard and similar to presentation software, this system allows for prior preparation of classroom materials, such as template documents that include animations and drawings, handouts, and quizzes. Unlike static presentation software but similar to a chalkboard, this system allows spontaneous responses, both from the instructor and from the students. However the additional capabilities are time saving. With a click of the instructor's mouse, the active application on any tablet in the classroom can be projected and, given permission, anyone can write on the projected tablet from his or her tablet. Electronic files can be sent to one or all tablets, surveys can be spontaneously generated (using the tablets as "clickers"), and electronic quizzes requiring drawings of complex chemical structures can be given. The instructor can lock out programs such as game software or the Internet, allow student chat, send and receive messages and take control of any tablet at any time. Using the recorder function a movie file can be made of any strokes made on the tablet, allowing a student or instructor to record a mechanism or equilibrium problem for later playback. Students can save their work to a server for later viewing.

POGIL in a Networked Tablet Classroom

The networked tablet classroom was used in POGIL chemistry classes at Eastfield College in Spring and Summer semesters in 2006. Although there were enough tablets for each student to use, Mewhinney chose to assign one tablet to each group of three, limiting the resource so that the students would be forced to work together. However, each student was given a tablet for individual electronic quizzes.

As class began, students used the tablets in the homework assignment discussion. A student with a homework question opened the OneNote™ software to the worked problem and the instructor projected it to the screen in the front of the room. Using the tablet, the structures did not have to be rewritten, and other students modified the answer using the tablet in front of them. A quiz was given after the homework discussion. On the occasions when an electronic quiz was given, each student received a tablet from the storage cart. At that time, any files needed for class were sent to their computers.

During most of the class the student groups worked the critical thinking questions from the OneNote™ copy of the graphics from the activity book *Organic Chemistry: A Guided Inquiry* (*7*), shown in Figure 4. Each student had a paper-based book in which they worked their own answers. The scribe for that day was responsible for recording the consensus answer on the tablet. The instructor observed the class's progress from her tablet and sent messages as needed. At logical points student work was projected to the front of the room and was discussed by the presenter in each group, and occasionally the instructor gave a mini-lecture. During class discussion the presenter in the original group, another group or the instructor amended the student answer

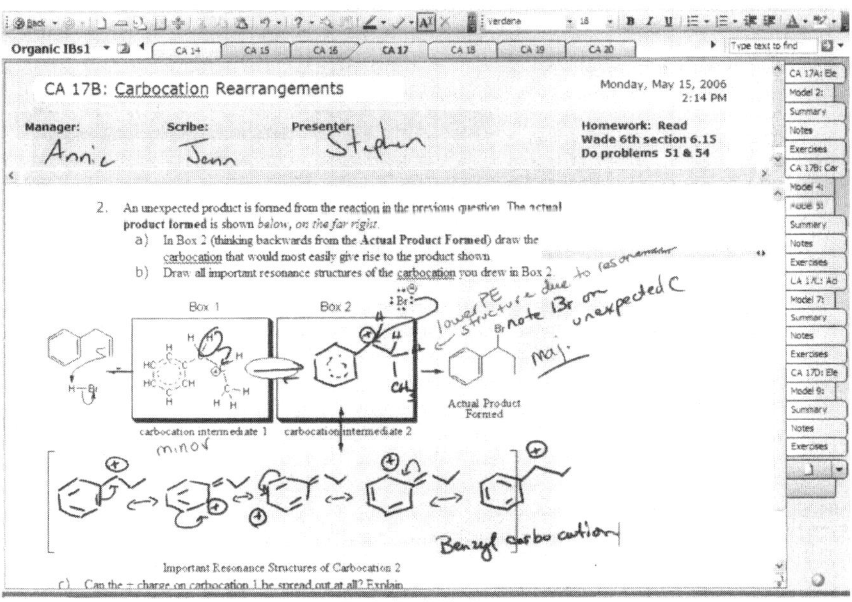

Figure 4. Student OneNote™ document complete with notebook organizational structure

projected. During these discussions the scribes were able to insert additional notes directly into their electronic document next to or on top of the model being discussed. Highlighting and drawing tools permitted notations, as one would do in the margin of a textbook, with the added benefit that any amount of space could be created wherever it was needed.

Impact on Learning

Since Mewhinney has used POGIL techniques in her classes for approximately 5 years, she was able to make some comparisons regarding teaching a POGIL class with and without this technology. Technology problems in the early networking phase detracted from the class. However, when the networking problems were resolved, the technology had a positive impact on the class. Several observations were made concerning the tablet classes versus POGIL classes in prior semesters.

Some student behaviors changed. With the tablet in front of them students demonstrated improved focus on the material and tended to stay on task for longer periods of time. In prior semesters students tended to work individually in their workbooks for several questions before discussing answers as a group.

With a single tablet as a focal point group members tended to stay in a discussion of one question until reaching a consensus before moving on to the next. In prior semesters, answers written in notebooks were often truncated and superficial, but, when using the tablets, scribes wrote more complete and in depth answers, as seen in Figure 5. Perhaps this was a result of not knowing when their work would be projected to the whole class (even anonymously), or perhaps because each scribe received a grade for the thoroughness of their answers based on the instructor's review of the group's work posted on the web site. More students were involved in their groups and more active in whole class discussions, and seemed to enjoy the class more, as seen in Figure 6. However, there did not appear to be increased interaction between groups, perhaps because of the lack of emphasis on this classroom capability by the instructor.

Managing the classroom efficiently and effectively was much easier for the instructor. Monitoring the groups' progress and determining the extent of individual student participation was relatively easy. In prior semesters "reporting out" of student answers and homework discussions was very time consuming as the presenter in each group had to rewrite the scribe's work on the board or on a transparency before reporting. This duplication was especially time consuming in organic classes where structures must be repeatedly redrawn. Using tablets, reporting out was much more time efficient because there was no need to recopy student work. The discussions were deeper and more students participated, perhaps because the projected answer was clearly written and visible. The student-archived work seemed to help students organize their notes, and was an effective tool for assessment of understanding and individual participation. All

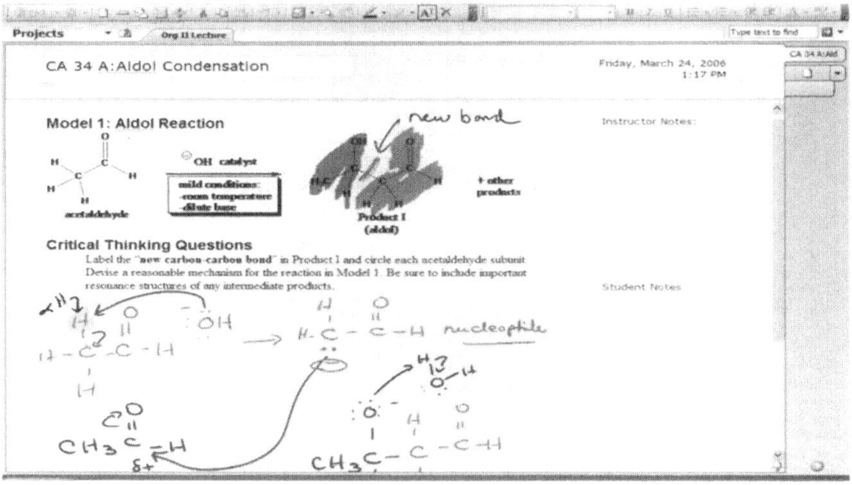

Figure 5. Student's answers written on a tablet were more complete

Figure 6. Projecting student work stimulated whole class discussion

in all, the interactions in the classroom were easier to manage and more time efficient, and more students were active participants in the process.

What was the POGIL tablet class like for the students? Members of the Summer 2006 pilot class responded to a survey (*8*) regarding their experiences. There were five response choices: strongly agree, agree, neutral, disagree, and strongly disagree. For simplicity, the two 'agree' categories were combined and the two 'disagree' categories were combined in the bar charts of Figure 7. Students were asked to agree or disagree with the following statements:

1. Writing on the tablet PC in my group helped me focus on the material.
2. Writing on the tablet PC helped my group stay on the same question until we agreed on an answer.
3. Writing on the tablet PC in my group helped us organize our class notes.
4. Seeing student's worked out solutions projected to the front of the room from their tablet aided in understanding whole class discussions.
5. Seeing the instructor's worked out solutions projected to the front of the room aided in understanding whole class discussions.
6. Using the networked tablet PC classroom made the class more interesting.
7. Using the networked tablet PC classroom helped me learn better.
8. It was easy to use the Microsoft OneNote software program.
9. It was easy to use the SMART Notebook software program.

The students were positive overall regarding the impact of the technology on their learning. There was a clear majority (if not unanimous) of Mewhinney's students who responded positively to all questions but two: writing on the tablet helped them stay on the question, and helped them focus on the material. The neutral responses to these two questions may reflect the technical difficulties the class experienced as the tablets were brought onto the network.

Flying Solo with a Tablet PC

It is unlikely that an institution's entire student body has tablet PCs to utilize in the classroom and for study. Until colleges and universities replace laptop and desktop computers with tablet PC's through attrition, the networked tablet technology as described above will be out of reach for most instructors. Still, a single tablet or a tablet monitor in the hands of an instructor can influence the classroom in many ways. Zückerman has chosen to employ a tablet PC due to the out-of-class advantages of the platform, such as note taking in meetings, portability, and constant access to all work files. The alternative to the tablet PC, a tablet monitor installed in a classroom, has the advantage of being available to any instructor using the room.

Zuckerman has utilized a tablet PC in his classroom for 4 years in classes following both traditional lecture and POGIL methods. In each scenario, one must now imagine a classroom with a single tablet, tethered to the front of the classroom via a cable connected to the projector—wireless connection to the projector is currently unavailable at Augusta State University (ASU).

POGIL in a Solo Tablet Classroom

One of the advantages to utilizing POGIL is the flexibility that exists as to how an instructor implements and facilitates their student's learning experience. At ASU, the tremendous variation in student skill sets and background knowledge must be considered when planning any lesson. Thus, depending on the concept and topic being covered, Zückerman's class utilizes POGIL as (1) a POGIL activity designed for the entire class time, (2) a 20-30 minute POGIL activity sandwiched between short 'mini-lectures' or other activities (such as a simulated experiment), or (3) several short 5-10 minute POGIL activities contained within lecture notes.

The seminal reason for the tablet in the small to medium classes at ASU is to ensure that the students leave the classroom with confidence that they have the correct answer in their activity. As the SALG data from Eastfield College displays, students consider the availability of correct answers vital to their success in the course. Prior to using the tablet PC, Zückerman's non-major physical science students did not employ the POGIL-like activities (*9*) when studying outside of class, citing inconsistencies in their answers (as compared to classmates from other groups) as the cause of their lack of confidence in the concepts portrayed in the activity.

As one might expect, there is a major pitfall to supplying answers to critical thinking questions in that some students or entire groups will do nothing until the answer is presented to the class and simply write down the answer provided. This is counterproductive, as the student is even less active than in a traditional lecture course. To combat scribing, Zückerman does not provide answers to any critical thinking question until each group has made an attempt at an answer.

Another advantage of utilizing the tablet PC in the POGIL classroom is that any discussion that arises in class (or at the start of class) can be compiled in one place, along with each activity and any accompanying notes. This may include email correspondence and office hours documents generated outside of class. Assessing the effectiveness of an activity or the response to questions is far easier when the instructor has all of the information in one or two files. The digital ink used to annotate files or in a digital whiteboard can be searched for keywords, facilitating faster research of ones topical coverage.

A typical day in Zückerman's POGIL classroom following option (2) from above begins with a short, traditional lecture of no more than 10 minutes. Notes for the class, which may include the POGIL activity, are made available online prior to class and are required for the student to participate in class. Upon completing the mini-lecture, a single focus question, designed as a 'big picture' conceptual question, is discussed in the groups. Whole class discussion of this question is done later. What follows is the POGIL activity. When all the groups have completed the activity, a few multiple choice questions designed to check for conceptual understanding are projected and discussed, first in groups and then by the class. Finally, the original focus question is discussed in groups. The class is then completed by either continuing with new material or moving

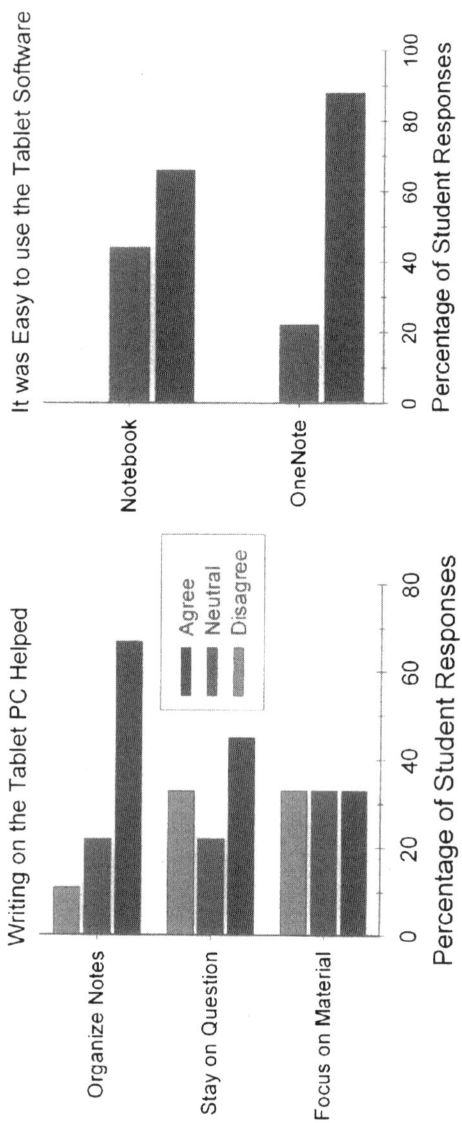
Writing on the Tablet PC Helped
Organize Notes
Stay on Question
Focus on Material
Agree
Neutral
Disagree
0
20
40
60
80
Percentage of Student Responses
It was Easy to use the Tablet Software
Notebook
OneNote
0
20
40
60
80
100
Percentage of Student Responses

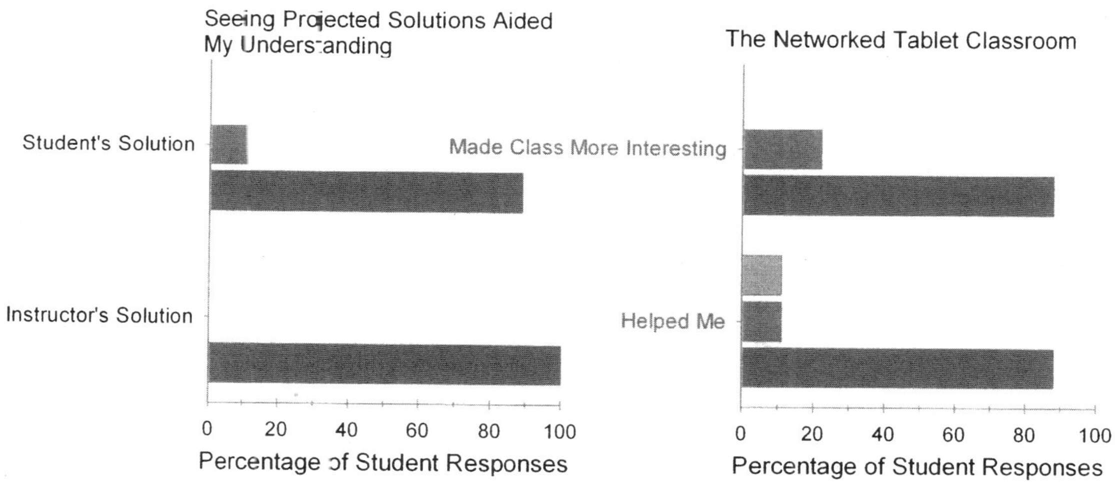

Figure 7. Results of Student Assessment of Learning Gains

more into applications of the concepts covered in the POGIL activity. Only rarely does a 50 minute class allow for multiple concepts to be covered in one period with multiple POGIL activities. Yet, it is commonplace for the concept and application portions of the activity to be broken up such that the time after the initial activity employs a second POGIL activity concerning with the mechanism of solving problems.

Often, science students are hesitant to trust instruction that is not presented in the traditional lecture format. More importantly, not all students are prepared for taking ownership of the learning process as freshman. For both reasons, implementation (3) is used to 'acclimate' students to the benefits of the POGIL activities during the first semester of general chemistry. After midterm, the focus shifts to implementation (2). The second semester of general chemistry is generally a mixture of implementations (1) and (2).

The tablet PC plays an integral role in the success of the above POGIL implementations. The use of digital ink allows the instructor to enhance any lecture portion by writing with color, highlighting regions of text, sketching diagrams and even opening a new blank page to diverge from the prepared materials. Software such as Classroom Presenter™ is designed to allow quick deviation from the prepared notes by either scaling down the projected image (leaving blank space surrounding class notes), or by generating on-demand digital whiteboards. The whiteboard ability of most ink enabled presentation software generates flexibility in the classroom that was otherwise unobtainable, especially in the large lecture format class. Instructors can now respond to student questions and focus class attention to the projected discussion not provided in the printed class notes. Additionally, instructors may generate 'on-the-fly' questions based upon the discussion within and amongst groups. The projection of the on-the-fly question ensures that each student has a proper copy (it is seen and heard) and, by the act of writing, the question is seen as relevant to the course.

The mini-lecture time may also be used in an alternative manner, such as using so called "mind mapping" software, as shown in Figure 8. Mind maps can be employed to capture student preconceptions about a topic or to visually organize relationships amongst concepts to be covered in a chapter, section, or

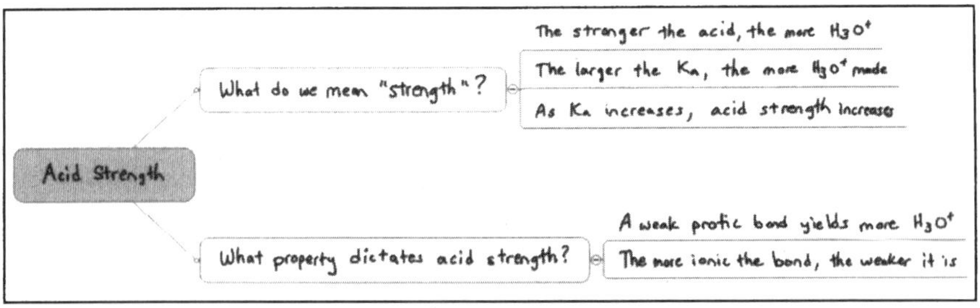

Figure 8. Portion of an instructor created Mindjet MindManager™ map generated during the mini-lecture portion of class

POGIL activity. As suggested by Knight (*10*), comparing concept maps based upon student beliefs to those created by the instructor is an awakening experience. By generating the student mind map directly from class discussion, the comparison of the structure and connections made by students and faculty can be appreciated by all in attendance. Thus far, mind mapping comparisons have been overwhelmingly welcomed by general chemistry students at ASU.

Impact on Learning

Although Zückerman has not utilized SALG surveys in his class, there are two important observations worth noting. Grades in his sections of introductory chemistry have improved as compared to the two years prior to introduction of both the tablet PC and the POGIL model, especially in terms of the percentage achieving an A or B. The drop rate, however, has reduced by approximately 10% over the same period.

In addition, student evaluations of the instructor have been more favorable. However, these same evaluations do not place the activities in as favorable a light. That is, some students do not see the POGIL activities as responsible for their success. Instead, a large number of students view their success as coming despite the POGIL activities. Students continuing their chemistry education do come to realize their improved long term learning and process skills, and often make this point known during advising sessions or informal meetings.

As in the introductory courses, the upper level physical chemistry course has seen a significant increase in performance and satisfaction with the professor since introducing the tablet PC and, especially, POGIL. But, unlike the introductory courses, these more seasoned students attribute their success to the guided inquiry methods. One student boasted that he "cannot imagine learning thermo by someone standing in front of me and lecturing."

Conclusion

The tablet PC can be a valuable tool in an active, student centered classroom. Networked student tablet PC's combined with classroom management software can enhance learning, and maximize the efficient use of class time from both an instructor's perspective and the student's perspective. The next step is the scale up of the technology to transform the static teacher-centered environment of a large lecture hall to an active, learner-centered environment, engaging students in a way as never before in a class of 500.

In addition, heretofore-untried pedagogical uses of networked tablet PC's should be explored. For example, tablet software, which records strokes made on a tablet PC and saves them to a movie file for later playback, could be invaluable in collecting student work that can be studied in an educational research environment. Studying the incremental output of the student brain, may lead to a deeper understanding of how students learn chemistry.

Networked tablets could also have a significant impact on distance learning by making it less isolated; small groups of the same class could meet in several different locations with an instructor who can interact with all of them. All in all it promises to be an exciting decade for the integration of technology in the active, learned-centered classroom.

Acknowledgements

Christina Mewhinney is very grateful to Hewlett Packard and Eastfield College for support of this work. This project could not have been successful without the outstanding and constant technical support from the IT department at Eastfield College. Eric Zückerman extends thanks to the Department of Chemistry and Physics at Augusta State University for their support in his applications of technology and POGIL. In addition, Eric is grateful to John Hill at Allegiance Technology Partners for creating the "48 Hour Tablet Demo" program that first introduced him to the benefits of the tablet PC. Both authors extend their thanks to the POGIL organization for extensive help with active, learner-centered pedagogy.

References

1. Wolfskill, Troy; Hanson, David. *J. Chem. Educ.* **2001**, *78*, 1417.
2. Farrell, John J.; Moog, Richard S.; Spencer, James N. *J. Chem. Educ.* **1999**, *76*, 570.
3. Hanson, David M.; Wolfskill, Troy. *J. Chem. Educ.* **2000**, *77*, 120.
4. Johnson, D. W.; Johnson, R. T.; Smith, K. A. *Active Learning: Cooperation in the College Classroom*; Interaction Book Company: Edina, MN, 1991.
5. *Process Oriented Guided Inquiry Learning, URL* http://www.pogil.org. Last accessed, October, 2007.
6. Mewhinney, Christina, *"High Tech-High Touch" Mobile Technology: Using Wireless Pen Based Technology to Enhance A Cooperative & Guided Inquiry Learning Environment in General and Organic Chemistry Lectures and Labs,* http://www.eastfieldcollege.com/smpe/chemistry/HPGrant/index.html. Last accessed, October, 2007
7. Straumanis, Andrei, *Organic Chemistry: A Guided Inquiry*, Houghton Mifflin: Boston, MA, 2004.
8. *Student Assessment of Learning Gains,* URL http://www.wcer.wisc.edu/salgains/instructor/SALGains.asp. Last accessed, October, 2007.
9. Adams, J.; Prather, E. E.; Slator, T.; Dostal, J. *Lecture Tutorials for Introductory Astronomy, 1st Ed.*; Prentice Hall: Upper Saddle River, NJ, 2005.
10. Knight, R. D., *Five Easy Lessons: Strategies for Successful Physics Teaching*; Addison Wesley: San Francisco, CA, 2004.

Chapter 15

Making Science Accessible in the Lives of Nonscience Majors Using POGIL and Project-Based Learning

A. Bryan Lees

Chemistry and Physics Department, Kean University, Union, NJ 07083

Engaging nonscience majors in chemically-related global and civic issues using rigorous chemical principles rather than descriptive discussion has been achieved using POGIL group classroom activities focused upon chemical principles, global issues and research skills. These activities provide an excellent foundation for developing in students the skills, confidence and comfort with scientific content to pursue independent literature projects on global and civic concerns. Based upon researched data, student project reports include a brief summary of an issue and one or more original calculations using conversion factors or stoichiometry to illustrate its magnitude, consequence, or resolution. Being able to answer a chemical question of their own making transforms the students' perceptions of their own abilities and of the accessibility of science in their lives.

Kean University has a 100-year commitment to transforming first-generation university students into successful professionals, originally serving the needs of the teaching profession as a normal school and now serving the needs of all academic, industrial and public institutions in New Jersey as a comprehensive, multicultural metropolitan teaching university. Its mission promotes access, excellence and a commitment to diversity in the composition of its student body and the academic preparation of its graduates. It has an enrollment of over 12,000 students, many of whom are poorly prepared or under-prepared in science and mathematics.

Comprehensive General Education curriculum innovations in 1999 at Kean University resulted from a planned and coordinated three-year effort to optimize the development and delivery of curriculum and academic support. It was in this context of General Education curriculum reform that Chemistry 1200, Chemistry in Your World, a one-semester, non-laboratory course for nonscience majors was approved with the intent of addressing science and math literacy in a framework of civic and global discussion. In order to support General Education oral communication skills, students are also required to give a 10-15 minute PowerPoint presentation on their project results in Chemistry 1200 at the end of the term. Since College Algebra is a course prerequisite, the only student group not well-represented is first semester freshman who are fulfillling this requirement. Otherwise, the composition of the class includes all undergraduates through graduating seniors, for many of whom science and math phobias have postponed their completing a General Education science elective. Class size is capped at 24 students, and two or three sections have been offered each term.

Only one course section per term has been offered in the POGIL-Project-Based Learning (PBL) format described herein, and because other sections have customarily been taught by a rotating staff of adjunct faculty, a consistent comparison of student performance has not been formally established between sections. However, the dramatic differences in student performance seen in my own POGIL sections are so compelling that they validate the methods described.

Pedagogical Challenges

Engaging nonscience majors in chemically-related global and civic issues using rigorous chemical principles rather than descriptive discussion is a realistic and rewarding goal. Yet, demonstrating student achievement in a one semester general education course has been for many faculty a long-standing challenge that has not been as successful as they may have desired. Classroom discussions generate short-term enthusiasm, but unless the entire class is focused upon a single issue, the depth of discussion may readily resort to a discussion about the sociology or politics of the issue and not provide an understanding of the underlying chemical principles that one hopes would be gained from a more rigorous course in chemistry.

Many textbooks for this scientific genre are oriented first to teach chemical principles in a traditional format and then apply those principles to chemical issues. With this approach, students spend most of their time in a lecture class where much of the transmitted chemical information is retained no more fully that it may have been in a previous high-school course. In fact, in such an environment, students often rationalize their frustration with the course as a result of their inability to remember "the answers" from high school rather than using reasoning and applying their analytical skills.

Using one of those few textbooks which incorporate chemical skills as they describe applied chemical topics, it is still difficult for the students to progress through more than a few topics within the time constraints of a single academic term. But, most significant to the educational growth of students, is that no matter how well written, or how integrated the topics, most of the information transmitted to students through most textbooks is expository. As a result, the ability of each student to assimilate new information and transform it into knowledge that they can fully appreciate and apply depends upon the previously developed cognitive skills that each brings with them to the class. Student outcomes then tend to preserve math and science phobias and the level of performance characteristic of each student's previous efforts, no matter how enthusiastic the students may be about a chemical topic they studied.

It was with this conundrum that I first attended a POGIL workshop in the summer of 2004. Having taught "Chemistry in Your World" in an interactive, but traditional, lecture format for one semester, I began to recognize and had come to welcome the following pedagogical challenges:

- establishing the mathematical and analytical skills needed for students to understand rigorous chemical principles rather than defering to more qualitative discussion;
- empowering ALL students through active reading, interpreting and solving chemical problems rather than reaching only the most motivated students;
- including stoichiometric calculations within the purview of student skills as the heart of chemical knowledge rather than resorting to only arithmetic conversions of simpler quantities;
- replacing purported disinterest in science with chemical skills that amaze students with their own abilities.

In spite of the daunting task, I left the workshop confident that I could incorporate guided-inquiry, small-group classroom activities, with independent student projects in a manner that would address the intended content, skill, and learning objectives of this course and well as my own pedagogical concerns.

Methods to Address Pedagogical Challenges: Combining POGIL with Project-Based Learning

Stoichiometric calculations would so expand the quantitative questions students could answer in their individual projects that its omission was not an option. Building guided-inquiry activites for the course was then defined by the need to reach this skill level as early as possible within the semester. The traditional textbook presentation of topics was reordered, simplified and conceptually integrated to support this fundamental objective. POGIL (*1*) classroom activities would be used to :

- build chemical and analytical skills,
- apply chemical concepts and stoichiometric calculations to global and civic issues, and
- solve and model open-ended problems that help prepare students for their own, quantitative, independent projects.

Project-based learning (*2-4*) through individual student projects would:

- summarize with pertinent chemical data the significance of a global or civic issue,
- outline chemical understanding of the issue using chemical equations, and
- answer each student's own quantitative questions that illustrate the magnitude of the issue, its effects, or its remedy.

Implementation Strategies for Using POGIL

The overriding challenge in using POGIL in this course is to provide students with the ability to read about chemical issues, to extract detailed chemical information, and to apply chemical skills to be able to complete their individual project. Overall, the course is divided into two components: the classroom which deals almost exclusively with group-based, POGIL skill-building activities, and homework which deals almost exclusively with research and writing for individual projects.

POGIL group activities are completed in 1-2 class periods of 1 hour and 20 minutes each. While no pre-lectures are given, brief mini-lectures are sometimes used in the middle or at the end of an activity to focus those students whose reading skills are weak and who have difficulty developing cumulative understanding. The need for such interventions varies considerably from class to class.

In the classroom, no long-term association of students in groups is established. Rather, students are randomly assigned to work in groups of four and the composition of each group changes after students complete each activity. Students are encouraged to work with every student in the class before working with the same student again. As a result, they quickly discern which other students work well with them, and often form out-of-class study groups with supportive peers.

The cohesive nature of POGIL activities and students' support for each others learning was found to depend heavily on the initial classroom implementation. A relaxed context for exploring and asking questions, and not for answering questions as quickly as possible, must pervade the intellectual climate for the entire term. Rereading the chemical content and requestioning each other leads students to far more successful performance outcomes.

Continual classroom experimentation by the instructor has led to a dramatic reordering of the chemical topics presented in classroom activities. ***Because a major course objective is to enable students to read about a wide variety of chemical issues as soon as possible, a conceptually-accelerated, but skill-appropriate method must be used.*** One example of this approach can be seen very early in the course. After completing an activity on the structure of atoms, ions, and free radicals, students are asked to recognize simple oxidation and reduction as the gain or loss of electrons from atoms and ions and to write simple chemical equations representing these processes. While limited in scope, this familiarity enables students to read about alternate energy resources such as batteries and fuel cells without being totally confused and without having to struggle through many other chemical concepts first.

All mathematical calculations, including stoichiometry, are treated as conversion factor problems and this recurrent approach provides reassurance to students with each new chemical concept. Even the most "math phobic" student can adapt to solving problems in this manner.

Outcomes Meet Pegagogical Challenges

After teaching this course using the combined POGIL and project-based learning formats for five successive terms, the effectiveness of POGIL activities in developing chemical skills is consistently compelling. On exams that parallel the critical thinking process in classroom activities, most students perform well, and approximately 15-20 percentage points higher than they did on more concrete exams in my interactive, but traditional, lecture class format. In addition at least half of the students also do well on unseen material. Including the first page of the next scheduled POGIL activity on an exam is frequently used as a measure of how well each student has advanced in reading and reasoning with new scientific material. Without question, POGIL classroom activities have enabled my students to master basic chemical skills more effectively, and, more importantly, ***to surpass their own expectations***.

Applying chemical skills in a quantiative independent chemical project is not only the most challenging course assignment, it ensures that students progress more fully into formal reasoning skills (*5-6*) and develop potential for becoming more critically constructive and scientifically literate citizens.

Completing their project calculations is usually more straightforward to many students, however, than focusing their project initially. When left to their own choosing, students often select a topic that is much too broad, much too comprehensive for them to be able to develop specific quantitative questions which they can answer. Kean students at all academic levels, including seniors, are prone to choosing a general topic such as "Environmental pollution" rather than a more focused topic such as, "The amount of limestone used to scrub sulfur dioxide from the smokestacks of coal-fired power plants." The project-based learning requirements for students to summarize with pertinent chemical *data* the significance of a global or civic issue and to outline chemical understanding of the issue using *chemical equations* have been found to be essential criteria for helping students focus their project questions. Using guided PBL combined with POGIL skill development in this manner results in increased student self-esteem that is unprecedented in the experience of this author. ***Being able to answer a chemical question of their own making transforms students' perceptions of their own abilities and of the accessibility of science in their lives.***

Impediments to Student Success

While many students adapt readily to the formats of this course, there is a significant number whose limited skill-level and classroom preconditioning greatly hinder their initial progress. While POGIL activities try to guide student learning in incremental steps that progress from simple to more encompassing concepts and conclusions, the ability to draw conclusions, itself, is for many students a challenging requirement. Drawing conclusions in a sequential process forces students to generate their own mental images that may be expanded and conjoined for cumulative understanding. No matter how incrementally their thoughts have been guided, however, those students who have not translated the original concepts into their own mental images are those who are easily "lost" and most resistant to POGIL. Contrary to my earlier expectations, the proportion of seniors struggling with this process is no less than the proportion of freshman.

As mentioned in a previous section on "Implementation Strategies," the process of students' constructing their own mental images cannot be rushed and must be painstakingly reinforced from the outset. Persistence must be encouraged and sufficient time must be allotted for students to read and reread and to recognize personally the potential reward of guided inquiry.

Guided Project-Based Learning

The structure of this course has a mixture of basic chemical skill-building and issue-oriented classroom activities predominantly in POGIL format. Ten of the fourteen activities given in Table I address chemical skill-building directly, three activites with bold-faced titles address chemical skills in the context of the global warming issue, and one activity with underlined title addresses quantitative research reasoning. It is Activity 2, *Chemical Issues and Quantitative Research Reasoning*, that has proven pivotal in guiding the focus of student projects. This activity "POGILizes" the quantitative reasoning skills requisite for a well-focused independent project. Its success suggests the applicability of POGIL to a wide-variety of project-based learning endeavors.

The placement of this activity very early in the course is also another critical factor in student learning. Contrary to many misconceptions, many students are extremely eager to understand the scope and breadth of a project announced on the course syllabus. Helping them to feel prepared to deal with the project as early as possible is necessary to dispel their apprehensions.

Activity 2 follows a POGIL format as outlined below.

Model of a Chemical Issue:

> An article in the newspaper states that "depletion of the ozone layer in the earth's atmosphere poses a serious threat to our lives because ozone protects both plants and animals on the earth's surface from too much damaging ultraviolet radiation from the sun."
>
> In order to understand this chemical issue, its causes, effects and remedy, you would need to answer a series of questions, many of them chemically related. Use your reference text (*8*) to answer the following questions that guide your understanding of this issue.

Note that the model directs students to use their reference textbook to answer the critical thinking questions that follow. Only in this activity, which is specifically intended to familiarize the students with the resources available in their book, are they encouraged to use reference material. All other skill-building POGIL activities are written to be self-contained, so that reasoned discussion, not "looking up answers" is the primary method of instruction.

The critical thinking questions that follow the above model require the student to look up information from several parts of the reference text, to distinguish ozone pollution in the troposphere from the ozone layer in the

Table I. POGIL Classroom Activities Addressing Chemical Skills, Global Issues, and Research Reasoning (*7*)

1. Elements & the Periodic Table
2. Chemical Issues and Quantitative Research Reasoning
3. Structure of Atoms, Ions and Free Radicals
4. **Climate of Man** *(9-11)*
5. Elements and Compounds
6. **The Greenhouse Effect and Carbon Dioxide**
7. **Carbon Dioxide Levels and Global Warming**
8. Unit Conversions for Solving Chemical Problems
9. The Mole Concept for Atoms, Ions and Formula Units
10. Conservation of Mass in Chemical Reactions
11. Stoichiometry
12. Solubility of Ionic Compounds
13. Oxidation-Reduction (Redox) Reactions
14. Understanding Electron Transfer in Electrochemical Cells

stratosphere, to understand the interaction of ultraviolet radiation with ozone molecules, and to identify the chemical cause of ozone depletion and the proposed remedies.

Quantitative research reasoning is addressed in the Problem section of the same activity.

Research Reasoning Problem:

The following statement appeared in the NY Times on Sunday, May 21, 2006 as part of an advertisement by Starbucks which "is committed to taking care of the world we live in."

"*If everyone who received this newspaper today switched one light bulb in their house to a compact fluorescent light, it would be like eliminating the emissions of approximately 89,000 cars for one year.*"

a. What type of emissions are referred to in the above statement?
b. How are those emissions related to automobiles and how could you determine their amounts?
c. How are those emissions related to the production of electricity and how could you determine their amounts?
d. Justifying the above quotation in detail would require you to answer a series of chemical questions quantitatively.
    i. What is meant by a quantitative answer to a chemical question?
    ii. List as many quantitative chemical questions as you can that would enable you to justify the quotation above.

NB: Completing and answering such a series of quantitative chemical questions would constitute an excellent class project.

Even though many students find working through Activity 2 very challenging on their own, they find solace in their group discussions and group research, and seem genuinely relieved to perceive what they believe to be their final course goal.

Their project efforts are divided into two reports, each of which is required to be complete. Never should the word "draft" be associated with any report. Previous conditioning leads most students to believe that a 'draft report' requires almost no effort and very little comprehension. To avoid such misconceptions, specific criteria for each report and the oral class presentation have been developed and are given in Table II.

Students who have not explicitly discussed their intent with the instructor will avoid addressing the criteria explicitly and will revert to a vague summary of the issue if given the opportunity. Such unsatisfactory behavior must be corrected with the first report if the student is to complete the course successfully.

Sample Project Calculations

Most students become comfortable with stoichiometric calculations through POGIL classroom activities and, once they are assured of the appropriate balanced chemical reactions, straightforwardly use them for their calculations. However, projects that rely only upon conversion factor relationships are equally acceptable as long as they are pertinent to the issue at hand and not inserted extraneously to satisfy the mathematical skills requirement of the course. One example of each type of project calculation is given below. The first uses numerous conversion factors to address the feasibility of corn-generated ethanol as a substitute for gasoline; the second uses a stoichiometric calculation of the effect of chlorofluorocarbon molecules on ozone depletion.

As one can see from the Sample Projects (Figures 1 and 2), student results can be both exciting and informative and the course can be most appropriately assessed by the number of WOW factors that result each semester. Many student results have not previously been recognized in either the popular or scientific press and for them to realize that their project is breaking new ground can be quite exhilarating. A particular case in point is Sample Project 1 which was completed in the spring 2006 semester. In August of the same year, a lengthy cover article, entitled, "The Ethanol Myth," appeard in Consumer Reports Magazine (*14*), reaching the same conclusions that were made in the student's project. Not only is the instructor impressed, but student self-esteem and enthusiasm with science is higher than has been observed in any other course format.

Conclusions

In surveys conducted at the end of the course, at least 80% of student responses are enthusiastic and comfortable with guided inquiry learning. Those students who are not comfortable are usually those who continue to insist on being given answers and resist individual and group reasoning. An important difference in student survey responses must also be noted. Student responses

Table II. Criteria for Completed Project Reports.

<u>**First Report**</u>:

a. A fully referenced, 3-5 page research summary of <u>data</u> that illusrates the significance of a chemical issue, written in expository prose.

b. Additional graphs, tables, and/or diagrams <u>with captions</u> are integrated within the research summary. Each item should be referenced.

c. A list of specific quantitative research questions that may be answered using the information, chemical equations, and other conversion factors included in the research summary.

<u>**Final Report**</u>:

a. A new version of the first report that includes corrections, additions, or revisions that were discussed with your instructor.

b. Specific quantitative research questions with answers and detailed calculations.

c. A discussion of the significance of the results.

<u>**Oral Class Presentation**</u>:

a. A title slide that focuses on your quantitative research questions.

b. A summary of the background and significance of the chemical issue.

c. A list of quantitative research questions that illustrate the magnitude of the issue, its effects or its remedy.

d. Answers to your research questions showing detailed calculations.

e. A brief summary of the significance of the calculations.

Part I: Net gasoline savings

An automobile driving an average of 12,500 miles/year with a fuel efficiency of 20 mpg would consume 625 gallons of gasoline or 925 gal of ethanol to obtain the same energy output. (from comparative enthalpies of combustion.)

To produce 925 gal of ethanol from corn, 24,100 lbs or corn or 3.40 acres of land are needed. (conversion factors from Cornell studies: 26.1 lbs corn/gal and 7110 lbs corn/acre.)

Approximately 140 gallons of gasoline are need to produce the fertilizer for 1 acre of land, so 476 gallons of gasoline are needed to produce enough corn to fuel one car with ethanol for one year.

Net gasoline saving: 625 gal - 476 gal = 149 gal or 23.8%

Student conclusion:

- Ethanol is not an efficient substitute for gasoline, but prolongs the available supply of petroleum.

Part II: Limits to ethanol production

Since 3.4 acres of land are needed to produce corn for 1 automobile each year, 680 million acres of land would be needed to supply the corn to fuel approximately 200,000,000 automobiles currently in the US.

Since the US now has a base of 470 million acres of arable land (decreasing at the rate of 2 million acres/year), less than 70% of the current ethanol demand could be met if every acre were devoted only to corn!

Student conclusions:

- Only a small fraction of our ethanol demand can be met with corn technology.
- A much smaller fraction of our gasoline demand can be saved with corn-generated ethanol.
- Alternate cellulosic ethanol technology should be investigated more thoroughly.

WOW!

Figure 1. Sample Project 1: The Feasibility of Corn-Generated Ethanol as a Substitute for Gasoline (12)

The mechanism for ozone depletion by a CFC molecule (Rowland and Molina) predicts that for every free chlorine atom released into the stratosphere, 100,000 molecules of ozone may be destroyed.

Assuming that a Cl atom is released from each freon-12 molecule, CCl_2F_2, 1 mol of freon-12 has the potential of destroying 100,000 mol of ozone.

If 1 refrigerator emits approximately 500 g of freon-12, a potential of 21.8 tons of ozone may be depleted for each refrigerator !

It is estimated that 8 million refrigerators are disposed of each year, so a potential of 74,400,000 tons of ozone may be depleted each year from refrigerators alone.

WOW!

Figure 2. Sample Project 2: The Effect of Chlorofluorocarons on Ozone Depletion (13)

tend to be much less instructor-oriented, unlike those is a more traditional lecture class. Rather than simply praising the instructor, student responses tend to be knowledge-, class-, and behavior-oriented and reflect the personal responsibility that each student has taken for his/her own learning.

The overall structure of the couse format proposed herein is not simply a merging of POGIL and project-based learning, but an extension of guided inquiry to all facets of student learning—chemical and mathematical skills, research, and global issues, as well as quantitative research reasoning. By guiding students in all areas of endeavor, their performance exceeds everyone's expectations and, especially, their own.

References

1. Farrel, J.J.; Moog, R.S.; Spencer, J.N. *J. Chem. Educ.* **1999**, *76*, 566-569.
2. Peterson, T. O. *J. Management Education.* **2004**, *28(5)*, 630-647.
3. Duch, B. J.; Groh, S.E.; Allen, D. E. *The Power of Problem-Based Learning*; Stylus: Sterling,VA, 2001.
4. Markham, T.; Larmer, J.; Ravitz, J. *Project Based Learning Handbook*; Buck Institute for Education: Novato, CA, 2003.
5. *Taxonomy of Education Objectives: Handbook I: Cognitive Domain*; Bloom, B.S., Ed.; David McKay Company, Inc: New York, 1956.
6. Perry, W.G., Jr. *Cognitive and Ethical Growth: The Making of Meaning* In *A. Chickering and Associates, The Modern American College*; Jossey-Bass: San Francisco, CA, 1981, pp 76-116.
7. Lees, A.B. *Chemistry in Your World: A Guide for Global and Civic Engagement*, 2005, unpublished.
8. Eubanks, L.P.; Middlecamp, C.H.; Pienta, N.J.; Heltzel, C.E.; Weaver, G.C. *Chemistry in Context, 5th Ed.*; McGraw Hill: New York, 2006.
9. Kolbert, E. *The New Yorker*. April 25, 2005, p 56.
10. Kolbert, E. *The New Yorker,* May 2, 2005, p 64.
11. Kolbert, E. *The New Yorker*, May 9, 2005, p 52.
12. Ballard, D. *The Feasibility of Corn-Generated Ethanol as a Substitute for Gasoline* In *Chemistry 1200: Chemistry in Your World*; Kean University, Union, NJ, 2006.
13. Cooney, C. *The Effect of Chlorofluorocarons on Ozone Depletion* In *Chemistry 1200: Chemistry in Your World;* Kean University, Union, NJ, 2005.
14. *The Ethanol Myth,* special issue cover title of *Consumer Reports,* October 2006.

Chapter 16

The Process-Oriented Guided Inquiry (Discovery) Laboratory

Ram S. Lamba[1,3] and Frank J. Creegan[2]

[1]Department of Chemistry, University of Puerto Rico at Cayey, Cayey, Puerto Rico 00736
[2]Department of Chemistry, Washington College, Chestertown, MD 21620
[3]Current Address: Chancellor, University of Puerto Rico at Cayey, Cayey, Puerto Rico 00736

A POGIL laboratory is one in which students, in advance of any classroom work on underlying principles, work in groups to conduct experiments, rather than exercises that verify previously taught principles. Prior to the beginning of any laboratory work the instructor poses a focus question (What factors affect the solubility of ions? When an alcohol reacts as a base, what role might substituents play?) and groups propose a set of tentative answers. To test these hypotheses, students run reactions and collect data, which are pooled and then analyzed with the aid of post-experiment or post-laboratory guided-inquiry questions. This learning cycle approach not only guides students to construct their own understanding of important chemical concepts but also helps them to develop valuable learning process skills.

The Guided Inquiry Laboratory Approach

The POGIL approach to the laboratory component is modeled on the guided-inquiry framework first introduced by Pavelich and Abraham (*1,2*), and popularized by Ditzler (*3,4*), Ricci (*5*) and the faculty at the College of the Holy

Cross as the Discovery Chemistry Approach. The essential factors of this instruction include emphasis on the processes and outcomes of scientific investigation as well as real problem solving through laboratory experiments that do not have a unique correct answer. The major goal is to involve students in the discovery of chemical principles from their own data in a process that illustrates the scientific method. To achieve this, students are involved in qualitative reasoning processes, modeled on the thinking processes followed by chemists and other scientists to organize data, develop principles, make predictions, formulate hypotheses and design experiments. Giving more emphasis to qualitative reasoning than to mathematical algorithms stimulates students to see the 'conceptual' aspect of chemistry from a perspective that is not present in more traditional approaches.

For over ten years we have been involved in the development, testing, and implementation of guided inquiry materials for general chemistry and organic chemistry courses. Student response to this approach has been highly favorable as evidenced by their comments in course evaluations and reflective diaries. Student success was measured by comparing the grade distribution in the guided inquiry-based courses, where the percentage of D-W-F grades was lower than in traditional courses. Among our students, this pedagogical approach has generated the enthusiasm and excitement for doing chemistry that characterizes our discipline and gives it intellectual vitality (*6,7*). In this chapter we describe two approaches to the use of the guided-inquiry method, one in which classroom and laboratory work are integrated and conducted on one space (also referred to as the studio approach) and a second in which the concepts developed by lab-based experimentation can be further discussed in a subsequent classroom setting.

The essential and desirable elements of our guided inquiry laboratory experiments are outlined as an unprioritized list of criteria in Figure 1. These have evolved from those developed by two groups of faculty: members of MADCP (Middle Atlantic Discovery Chemistry Project), a consortium of thirteen schools in the Middle Atlantic region, which was created in 1993 and which received financial support from the US Department of Education's FIPSE (Fund for the Improvement of Post Secondary Education) Program; and faculty involved in a series of NSF-funded and MSEIP-funded programs in Puerto Rico.

Two criteria require special emphasis. The POGIL approach is one of guided inquiry and not open inquiry. It is vitally important that the instructor, acting as facilitator, know in advance the outcomes of the experiments. The experiments must be tested so as to produce reliable data, which when pooled allow students to develop chemical concepts.

Typically, a laboratory session starts with a discussion session in order to focus the specific laboratory exercise. Students, organized in groups, are asked a focus question (note the first criterion in Figure 1) to formulate hypotheses, which are then discussed among the group members. In some cases, students are

Begins with a question.
Title does not reveal the concept to be discovered
Involves hypothesis formulation
Uses observation or data collection to develop theoretical construction rather than confirming a concept.
Keeps instructor input into experimental procedures at a minimum.
Promotes active decision-making.
Prior to experiment, outcome is known to instructor but not to student.
Designed so that students can get reliable data.
Uses critical thinking questions as well as student-student and instructor-student interactions to guide students to the appropriate conclusion.
Guides the student in recognizing what has and has not been learned from the experiment through use of appropriate in-lab and post-lab questions.
Reinforces the developed concept through application.
Promotes teamwork
Encourages students to develop questions for further research.
In some cases allows student input in design of experiments.

Figure 1. Criteria for POGIL experiments

asked to make predictions about the data they expect to collect. The experimental part in each laboratory experiment is characterized by a division of labor. Each group of students is assigned a variation of the data to be collected in a particular experiment. For example, in a kinetics experiment a few groups of students work on one set of concentrations and temperatures, while the other groups work on different sets of concentrations and temperatures, and so on. At the end of the experiment, all groups share results. The combined data are utilized for data interpretation and concept development. Furthermore, the use of combined data permits student discussions in small groups as well as among the whole class, which allows for the exchange of ideas, verification or rejection of hypotheses, and discussion of the implications of the collected data. The combined data are utilized to make graphs and tables, and to answer questions in order to derive the concept under study. Our goal is to provide students both with a more accurate picture of the scientific process and also the qualitative problem solving skills that are relevant to solving real life problems. This format is summarized in Figure 2.

Pre-lab Session	*Experiment*	*Post-Experiment*
• Present focus question • Solicit hypotheses or predictions • Clarify questions about experiment and experimental setup	• Each team is assigned a variation • Pool data • Discern trends	• Graphical analysis • Interpretation of data • Discovery of concept

Figure 2. Laboratory format

Parallels Between Piaget's Theory, the Learning Cycle and Guided Inquiry-Based Laboratories

There are strong parallels between Piaget's Theory (*8,9*), the Learning Cycle (*10*) and this approach to laboratory instruction. These parallels are illustrated in Figure 3.

Recently Abraham (*11*) has described in some detail how the assimilation-accommodation-organization processes as described by Piaget are similar to the exploration-concept invention-applications phases in the Learning Cycle. In POGIL laboratories, students collect and analyze data, an exploration phase that allows them to transform or assimilate information into their existing mental structures, which in turn allows them to accommodate to it. Having integrated the new information with existing mental structures, students can meaningfully interpret their experiences to derive concepts, which can be verified or modified through applications and extensions where questions related to the previously learned concepts are presented in new contexts. This step is important to insure students have truly mastered an understanding of the developed concepts.

The Learning Cycle has three instructional phases in a specific sequence: E - I - A. In the first two phases, the students use the data collected in experiments to invent the concept inductively; in our model students generate data and answer questions in order to derive the concept to be developed. In that sense, this is one of the most important phases of the learning environment. In the A phase the students verify, modify, or extend their ideas. In this respect, this phase is a deductive process. Course evaluations and grade distributions in our courses are consistent with the observations of Michael Abraham and others (*12,13*), who report that the learning cycle approach in general results in greater achievement in science, better retention of concepts, improved attitudes towards

Piaget's model	*The Learning Cycle*	*Guided Inquiry-Based laboratories*
Assimilation	Exploration (E)	Data analysis and implications
Accommodation	Invention of Concept (I)	Interpretation of data and conclusions
Organization	Application (A)	Extension and Applications

Figure 3. Piaget's Theory, the Learning Cycle and the Inquiry-Based Laboratories

science and science learning, and superior reasoning abilities and process skills, as compared to more traditional instructional approaches.

Guided Inquiry Experiments for General Chemistry: Practical Problems and Applications

Guided Inquiry Experiments for General Chemistry: Practical Problems and Applications, 1st Edition is a laboratory manual (*14*) that is designed for students in introductory chemistry courses. The product is the result of several projects funded by the National Science Foundation and the US Department of Education-MSEIP. The experiments in this manual have been designed to enhance students' thinking skills and understanding of key chemical concepts within a practical problem-solving context. For example, each experiment begins with a problem scenario ("Why Did My Watch Stop Suddenly?"; "How Deep Can a Diver Go?"…) and ends with questions requiring feedback on the problem. An objective is to develop students' thinking via inquiry with an emphasis on hypothesis testing (what do the observations mean?), as well as analysis and visualization skills. A goal is for students to learn how to approach and solve problems (what data should you measure, what samples should you use, what does the data indicate? etc.) while gaining appreciation for the practical applications of chemistry. The concluding section ("Extension and Applications") of each experiment poses questions that require students to extend findings to untested systems and real world situations.

Design of Experiments

The design of the experiments is based on the constructivist theory that students construct their knowledge from involvement, experience, and models.

Student dialogue and collaboration is encouraged so as to maximize the contributions peer collaboration can make in improving understanding. Information, models, and guidelines are provided (rather than explicit "recipes" or detailed procedures) to encourage active student involvement. Students, however, are not expected to know or "construct" skills or techniques. For example, experiments in the introductory unit ("Scientific Method") provide explicit directives and models for basic skills such as designing tables for data collection and drawing graphs. Details are streamlined in subsequent experiments to promote active learning.

Sufficient information and guided inquiry questions within every experiment for both safe and successful learning outcomes are included. A Teacher's Guide has been developed for each experiment, providing the key concepts the experiment develops, skills used, materials and chemicals needed, anticipated data, ideas for classroom implementation, and sample answers to questions.

The guided inquiry experiments are sorted into 12 units based on topics common to most general chemistry texts. Each unit offers several experiments on a given topic, each focusing on a different key chemical concept. For example, Unit 4, Reactivity and Periodicity, has two experiments: "What factors affect the solubility of ions?" and "Can toxic ions be removed from water by precipitation?" Similarly, Unit 9 on Acids and Bases includes two experiments: "Is it acidic, basic or neutral?" and "Are Acid-Base Properties Predictable in Consumer Products?" Experiments are arranged sequentially within a unit so as to build upon one another. Experiments that occur later within a unit or later within the text can be seen in some instances to be extensions or applications of concepts developed earlier.

Instructional Model

In our experiments, in which the data are collected for a variety of reagents, students are more readily convinced that they are observing general trends rather than unique examples. Thus, if a single group observes that addition of $CoCl_2$ to water causes a decrease in pH, they may conclude that color and pH are correlated (this fits the students' model of pH indicators). If every group observes that addition of transition metal salts causes a drop in pH, students are more likely to conclude that transition metal aquo-ions are acidic. From these data, students can readily observe periodic trends in pH, solubility, and complexation. The success of pooled data in supporting learning is consistent with the findings of Johnstone and others (*15*) that students must be put in situations in which they explore the similarity in effect for an apparent diversity of instances.

The instructional model allows us to emphasize scientific thinking and associated cognitive skills. Students are asked to organize the class data in order to determine if there is a link between the observed properties and structure: "Does acidity correlate with the cation's ionic radius (or charge, or placement of the ion's element in the Periodic Table...)"? Students are able to discuss and analyze questions which require them to apply science in the real world. For example: "A waste site contains KNO_3, $Ca(NO_3)_2$, $Cu(NO_3)_2$, and $Pb(NO_3)_2$. The ground water is found to be acidic. Which (if any) of these salts could account for the acidity? Refer to the class data to support your conclusion."

This type of experimentation directly illustrates the variation and uncertainty of science, since everyone's results don't agree. Since students are able to (as a class) collect data under a variety of conditions, and/or with a variety of reagents, they are better prepared to make predictions about unfamiliar situations (i.e., to "think like a chemist").

Following the activity, the students assemble for a post-lab session in which they discuss the group results which can lead them to discover a wide range of new concepts, as well as to answer the original question. The discussion may continue through several additional class meetings. A typical experiment is described below.

How Long Can a Bubble Last?

Typically, an experiment begins with a question. In this case, the experiment's objectives are to investigate the effects of varying conditions (soap concentration, its temperature, bubble size) on the time that a bubble lasts (*7,14*). In addition, students are asked to design an experiment for determining the effect of different surfaces on bubble longevity.

Students are provided with four soap solutions of different concentrations and asked to practice with one of the solutions to blow bubbles. They are asked to predict what should happen to the lifetime of the bubble as the concentration of the soap is increased. (Most students predict that as the concentration of the soap solution increases, the time taken for the bubble to burst [its lifetime] also increases.)

Then, instructions call for blowing bubbles of a known diameter on a flat laboratory surface using 1 ml of a liquid soap solution (Dawn® works best in our experience). They work in groups of two and record the instant the bubble bursts. They repeat their observations with all four concentrations until they get consistent results.

Each group collects its data and then all data are displayed for discussion. Each group's predictions are compared with the class data. As part of the objectives, the students are asked to present data as a graph of the time of bubble bursting vs. concentration of the soap solution, as shown in Figure 4.

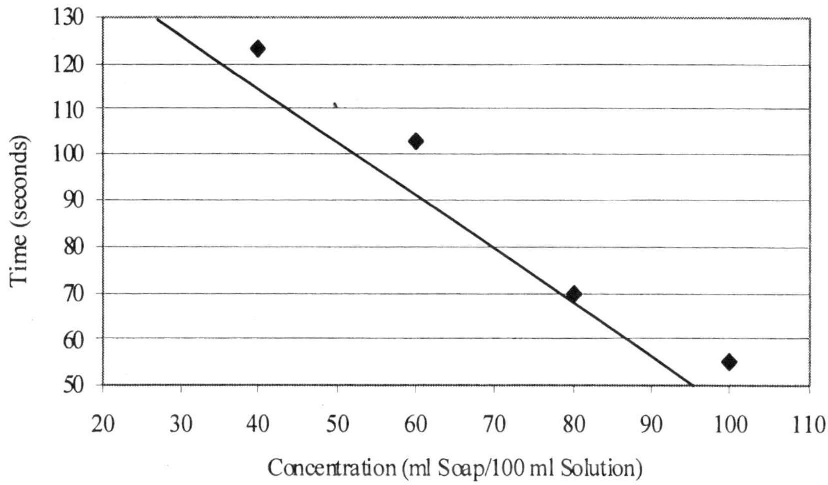

Figure 4. Effect of the concentration of the soap solution vs. time

The teacher leads the discussion as to why their prediction and their observations contradict. Factors on which the lifetime of the bubble depends, such as hydrogen bonding, cohesive and adhesive forces, are brought in through their discussion and observations.

Then the students are asked to hypothesize what will happen to the lifetime of a bubble as the temperature varies. (Invariably, they hypothesize that as the temperature increases, the lifetime of the bubble decreases.)

They are divided into groups and make their observations at three different temperatures such as, 10°C, 20°C and 25°C. That is, one group of students perform the experiment at 10°C the other at 25 °C, and so on.

The students are asked to provide their own data table and plot a graph of the time of bubble bursting vs. temperature of the soap solution for a specific size of the bubble. Again, their observations are not consistent with their hypotheses. The class data are again displayed in the form of a graph.

Finally, the students are asked to design an experiment(s) to study the "effect of the size of the bubble," and the "effect of different surfaces" on the lifetime of the bubble. In each part, in the results and conclusion section, they are asked to summarize their findings and give a possible explanation for the problem under investigation. Through this activity, students learn about scientific method and discover concepts such as adhesion, cohesion, the wetting of a solid surface by a liquid and factors affecting reaction rates. Hydrogen bonding is reinforced and discussed further.

Guided Inquiry Experiments for General Chemistry: Assessment

We have used the Guided Inquiry laboratories since 1995 with great success. The first time around, external evaluators provided us with this evidence of success, not only by comparing the grade distribution in the guided inquiry-based laboratories with the traditional courses, but also by measuring the improvement in student attitudes towards studying chemistry, which was linked to greater confidence in their ability to perform adequately in such courses. They also collected qualitative data using Reflective Diaries that students completed regularly and interviews during the course.

Due to the fact that the guided inquiry laboratory course is integrated with no distinction made between or definition given to lecture and laboratory, and that we have made the laboratory the centerpiece of the students' learning experience by introducing concepts by laboratory experiments (*16*), the evaluation studies of the method were conducted in the laboratory setting. The course used the guided inquiry approach activities developed by Moog and Farrell (*17*). The students earned the same grade for the lecture part of the course as for the laboratory component. In the traditional course, in most instances different faculty members give the lecture and laboratory components, and there is basically no integration as reflected by the fact that in the traditional course most students earn a much better grade in the laboratory component than in the lecture component.

Quantitative measures included a comparison of grades obtained by students taking the guided inquiry course and that of the traditional course and laboratories. The Chi-square comparison of traditional and guided inquiry formats consistently revealed a higher percentage of B's and C's in the guided inquiry approach and a lower percentage of withdrawals and F's. No significant difference was found in terms of A's between the two courses. Thus, a group of students (D, F and W) moved towards C, whereas students obtaining a grade of C moved towards B in the guided inquiry course.

In order to determine the retention of concepts learned during the introductory chemistry course taught using the guided inquiry approach, a follow-up study was made of their grades in subsequent chemistry courses, such as Organic Chemistry. Although the sample is small, as the follow-up study consisted of one academic year, it was clear that the students who took the guided inquiry approach obtained a higher percentage of A's and C's and a lower percentage of D's and F's as compared with those of the traditional approach.

Student Reflections

The use of Reflective Diaries provided qualitative data regarding the course and the laboratory. These diaries were designed by the external evaluators and were analyzed monthly. Anonymous feedback was provided regularly to the professor. The diaries included topics such as: more difficult concepts, difficulties in the laboratories, skills which students needed to practice more, concepts and skills that were easy for them to learn, areas of achievement, recommendations for the professor and their general perception of the course. The anonymous feedback indicated where additional discussion on some concepts and additional practice in some skills would be helpful. Some quotes from the diaries written by the students follow:

"The guided inquiry course is wonderful, you learn a lot and you know how things happen and why do they happen. The class is based on experiments and you learn a lot more by observing what you are doing than by listening to a professor."

"The system allows for you to obtain knowledge by yourself through the experiments, you do not have to rely only on textbooks."

"Allow me to make you an invitation to the world of chemistry. This class is like a window to a world that you could not think exists. It is a way of showing you that chemistry is something that has to do with you in your daily life, even if you had not realized it before. And it is in this course that you realize this. It is a totally progressive method, and you learn the concepts by doing, that are almost impossible to master well unless you take the guided inquiry course."

Guided Inquiry Experiments for Organic Chemistry

At Washington College a full year of POGIL experiments for the Organic chemistry course has been developed, tested, and implemented. Each follows the learning cycle paradigm that Abraham and Pavelich (*18,19*) first brought to the undergraduate chemistry laboratory curriculum in 1979 with their pioneering text, *Inquiries into Chemistry*. Illustrative of this approach is an experiment, which works well in General Chemistry as well as an introduction to the sophomore-level Organic chemistry laboratory, in which the instructor asks, *How is the structure of a molecule related to its boiling point?* Groups, with only the limited information provided by the names and boiling points of two simple organic substances, are asked to list all the factors related to molecular structure that might influence the boiling point. The groups are then given a set of four liquids and asked to measure the boiling points. The data are collected.

First using the group data and then by using the class data students test their hypotheses. Several of the groups are assigned a set of homologous alkanes, or primary alcohols, or secondary alcohols, or ketones, or alkyl bromides, but one or more is assigned a set composed of representatives from each of the five classes. In examining the data from a homologous series, such as that shown in Figure 5, students quickly see that molecular weight is an important factor. However it is in the pooling of the class data that students come to see that additional factors need to be considered in order to explain how a ketone and an alkane with nearly identical molecular weights (shown in the vertical rectangle in Figure 5) have considerably different boiling points (2-hexanone, MW=100.16; BP=127°C and n-heptane, MW 100.20; BP=98°C). Such examples of discrepant behavior provide for lively discussions. Student-generated boiling point data are coupled with modeling data on dipole moments, atom charges and electron density surface images, which lead to the development of an understanding of the factors that contribute to intermolecular forces.

In part 2 of the experiment a similar focus question on melting points is presented. Students are quick to suggest the role of molecular weight and the presence or absence of functional groups as key contributors to melting point. Students use a MelTemp apparatus to obtain melting point data and Spartan Student Edition to construct pdb images of their samples that are examined with RasMol-UCB. Images such as those shown in Figure 6 for anthracene (MP=214) and 9,10-dihydroanthracene (MP 108) help students to come to an understanding of the role that steric factors play in physical properties of organic molecules.

In the next experiment, which might form an extension phase to the developed concepts, students are presented with a new focus question, *"How and why do substances dissolve?"* Answers are generated from solubility date obtained from testing sets of homologous and isomeric alcohols in water and in a set of hydrocarbons (hexane, octane, decane). Later in the course students are asked to cycle back to these concepts as they conduct experiments designed to uncover organic reaction mechanisms.

Illustrative of this is an experiment conducted midway in the first semester, in which students are asked; *When an alcohol reacts as a base what role might substituents play?* To test their hypotheses a set of isomeric C-6 alcohols are treated with 85% phosphoric acid and in turn heated under reflux during which low-boiling material is removed by distillation. Students are aware that the alcohols under investigation have boiling points that range from 130°-150°C. In a set of prelab questions students are encouraged to recall, by returning to data from the first experiment of the semester on boiling points, the factors that might be responsible for these observations. In the Exploration Phase students note that the boiling points of the distillates are nearly half that of the alcohol substrate, and that tertiary alcohols produce the low-boiling distillate (react) at a much faster rate than do the secondary alcohols, which react much faster than the

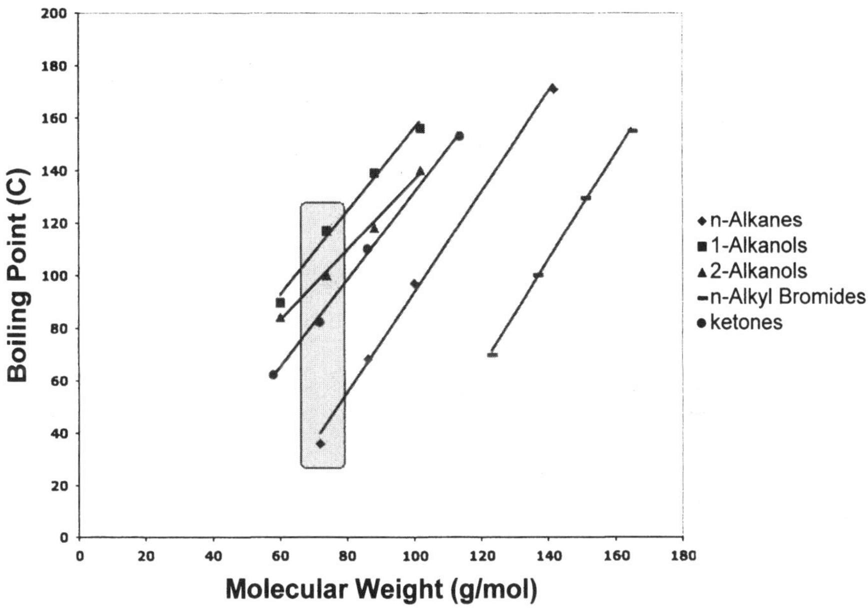

Figure 5. Boiling Point Trends

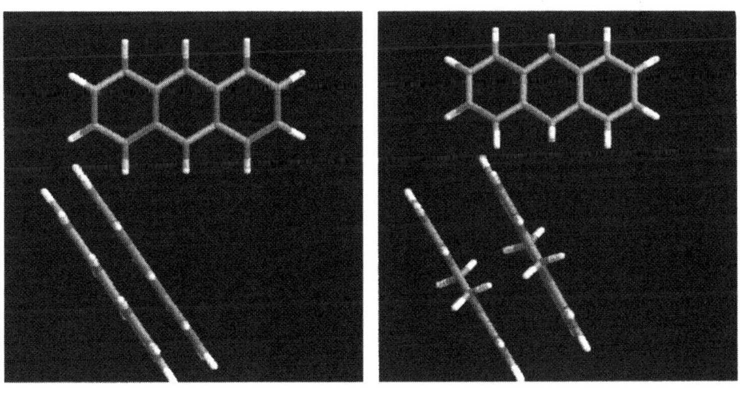

Figure 6. Views of anthracene (left) and 9,10-dihydroanthracene (right)

primary alcohols. Students examine the distillates by GC or GC/MS and discover the products to be mixtures of isomeric, low boiling alkenes. The pooled chromatographic data reveal structural features that are best explained by invoking skeleton rearrangements arising from a mechanism that involves carbocation intermediates. Rate data, product information, and a set of well-scaffolded post-laboratory questions allow students to develop the E_1 mechanism for acid-catalyzed dehydrations.

Conclusions

In the POGIL laboratory students make their own hypotheses, test them through their experimentations and observations, and analyze and interpret the data with guidance from the instructor and in collaboration with their peers. There are sufficient elements of creativity developed in this format to support the insights expected of the students in the process of discovery. Throughout the process, the instructor is behind the scenes, a guiding force, choreographing the activities, insuring that student creativity is called for in small, manageable increments.

Our small sample follow-up study of students in subsequent courses shows better retention of concepts, improved chemistry learning, more self-confidence and improved research skills developed with the guided inquiry modality.

References

1. Pavelich, M.J.; Abraham, M.R. *J. Coll. Sci. Teach,* **1977**, *7(1)*, 23-26;
2. Pavelich, M.J.; Abraham, M.R. *J. Chem. Educ.* **1979**, *56*, 100-103.
3. Ditzler, M.A.; Ricci, R.W. *J. Chem. Educ.* **1991**, *68*, 228-232.
4. Ditzler, M.A.; Ricci, R.W. *J. Chem. Educ.* **1994**, *71*, 685-688.
5. Ricci, R.W.; Ditzler, M.A.; Jarret, R.; McMaster, P.; and Herrick, R. *J. Chem. Educ.* **1994**, *71*, 404-405.
6. Lamba, R.S.; Lloyd, B.W. *J. Chem. Educ.* **1997**, *74*, 1095.
7. Lamba, R.S. *The Journal of Mathematics and Science* **2001**, *4*, 115-125.
8. Piaget, J. *The Origins of Intelligence in Children;* Norton: New York, 1963.
9. Piaget, J. *Structuralism;* Harper and Row: New York, 1970.
10. Karplus, R.; Their, H.D. *A New Look at Elementary School Science*; Rand McNally: Chicago, 1967.
11. Abraham, M.R. *Inquiry and the Learning Cycle Approach.* In *Chemists' Guide to Effective Teaching*; Pienta, N.J.; Cooper, M.M.; Greenbowe, T.J., Eds.; Prentice Hall: Upper Saddle River, NJ, 2005; pp 41-52.

12. Lawson, A.E.; Abraham, M.R.; Renner, J.W. *A Theory of Instruction: Using the Learning Cycle to Teach Science Concepts and Thinking Skills;* Monograph Number One; National Association for Research in Science Teaching, Kansas State University: Manhattan, KS, 1989.
13. Lawson, A.E. *Science Teaching and the Development of Thinking;* Wadsworth Publishing Company: Belmont, CA, 1995
14. Lamba. R.S.; Kerner, N.K. *Guided Inquiry Experiments for General Chemistry: Practical Problems and Applications, 1st Edition;* John Wiley and Sons, Inc.: New York, 2008.
15. Johnstone, A.H.; Letton, K.M. *Education in Chemistry.* **1991,** *28,* 81-83.
16. Lamba, R.S. *J. Chem. Educ.* **1994,** *71,* 1073-1074.
17. Moog, R.S.; Farrell, J.J. *Chemistry: A Guided Approach;* 3rd ed.; John Wiley and Sons, Inc.: New York, 2005.
18. Abraham, M.R.; Pavelich, M.J. *Inquiries into Chemistry,* 3rd ed.; Waveland Press: Prospect Heights, IL, 2000.
19. Abraham, M.R. and Pavelich, M.J. *Inquiries into Chemistry,* 4th ed.; Waveland Press: Prospect Heights, IL, 2004.

Chapter 17

Implementing POGIL in a Multiple Section Laboratory Course

Scott E. Van Bramer and Andrea E. Martin

Department of Chemistry, Widener University, One University Place, Chester, PA 19013

Implementing significant changes in multi-section laboratory courses is a complex task. This chapter outlines some steps that were important to the revision of our general chemistry laboratory sequence. It should prove useful to departments that are considering implementation of POGIL in multi-section laboratory courses.

The complexity of implementing a new model for laboratory instruction is often a significant barrier for courses with multiple sections taught by multiple instructors. Overcoming these barriers and successfuly catalyzing change in the curriculum requires time and planning. The chemistry department at Widener University has spent the last three years revising the general chemistry laboratory sequence. We hope the process that was used to implement these changes at Widener University will help others who are considering similar changes at their institutions.

The Department of Chemistry at Widener University offers a two semester general chemistry sequence for science and engineering students. The sequence is taken by approximately 120 students each year. Typically about 75% of the students are engineers and 25% are science majors – the majority of whom are in biology. The first semester we run four or five lecture sections and eight to eleven lab sections. The lectures are all taught by full time tenure-track faculty. Some of the laboratory sections are taught by the faculty and some are taught by part-time adjunct faculty. The department had been using a very traditional in-house laboratory manual. Over the past four years the department has developed and implemented a new laboratory curriculum that has significantly changed how we teach general chemistry. Careful planning of the process used to implement this change has been important in the success during the implementation. The first step of this change required several catalysts to start the process and increase faculty buy-in for change.

Catalysts for Change

One of the most important catalysts was the emphasis on assessment brought about by an upcoming Middle States accreditation visit. In the 2002/2003 academic year, departments in the College of Arts and Science began discussing assessment for the first time. One of the first assessments done by the chemistry department was a survey of faculty outside of chemistry to identify how well our courses were meeting their students' needs. This information, combined with the results from course evaluations, pointed out some problems with our previous laboratory course design. The comment "I would like to see more open ended projects rather than over-reliance on a completely scripted set of cookbook lab activities" was consistent with many faculty and student attitudes about the original course. As a part of this process, the chemistry faculty also spent a considerable amount of time discussing what we wanted students to learn in chemistry courses. During these discussions it became clear that the laboratory experience we were providing was not designed to meet our objectives for our students.

At the same time faculty in the chemistry department were attending conferences and workshops that challenged the traditional structure for laboratory courses and questioned the traditional priorities for what students need to learn. The University provided travel funding so that most of the faculty in the department could attend these meetings. They included a Project Kaleidoscope workshop "Shaping General Education Programs Focused on Scientific & Quantitative Literacy" in New York City in November 2003, the Middle Atlantic Discover Chemistry Project meeting in Washington DC in June 2004, and the Biennial Conference on Chemical Education in Ames, Iowa in July 2004. In these meetings the faculty were exposed to new ideas about how chemistry laboratories could be taught. The meetings also provided an opportunity for the chemistry faculty to spend a significant amount of time talking about teaching. The conversations were much more focused and thoughtful than is possible during a short conversation in the hallway or the first ten minutes of a departmental meeting.

Another important catalyst for change was the POGIL project. A group of Widener faculty visited Franklin and Marshall for a day to see how POGIL is implemented in classrooms. A month later one of the POGIL Project leaders spent a day at Widener doing a POGIL consultant visit. During these visits faculty had an opportunity to see how POGIL works, ask questions about implementing change, and get some personal feedback about their teaching.

These catalysts combined to provide an opportunity to make substantial change in the chemistry curriculum. Based upon our experience, any department that is considering large-scale change should try to use as many catalysts as possible. Program assessment provided a forum for the faculty to discuss the goals and objectives for our courses and to realize that our courses were not structured to meet those goals. It set the stage where faculty readily accepted that what we were doing needed to change. Assessment is critical to getting buy-in for change. Assessment also provides a framework to evaluate the effects of change and to continue to improve how we teach chemistry. In hindsight, we should have spent more time as a group working out the goals and objectives for the classes. Many of the conflicts that we did have were about which goals are most important. What we found is that there are a tremendous number of skills that we want students to learn from an introductory laboratory. It is probably not possible to address all of them fully, so some compromise is necessary. Having these conversations at the beginning of the process provides a framework to make decisions about what to spend time on and what to grade. Attending conferences and the POGIL visits were also critical because they provided us time to talk and reflect. These opportunities helped the department come to a consensus about what we wanted to try and why it was important. It also allowed us to learn about concerns and reservations that some of the faculty had about change. This communication has been critical to implementing change.

Plan for Change

Once the faculty had decided that we needed to change, we started to work out a plan for how to implement the changes. We spent a significant amount of time discussing different ideas and strategies to find a way that would address faculty concerns and reservations. In hindsight, this time was very well spent. The strategy we developed was critical to successfully implementing large scale change in our laboratory curriculum. The key pieces of our plan included:

1. Starting with the first semester in the general chemistry sequence
2. Obtaining a summer salary stipend to spend time rewriting the first semester laboratory manual – Summer 2004
3. Running a pilot using a single honors section – Fall 2004
4. Running the lab off-sequence with a small number of students and only two instructors – Spring 2005
5. Rolling out to multiple sections – Fall 2005

We decided to start with the first semester of the general chemistry sequence because it had an impact on a large number of students and provided an opportunity to make changes before students develop expectations of what college chemistry lab is like. Based upon our assessment, this was the course that was turning students off to chemistry. This was the course that regularly received comments like "I like chemistry, but I hate the lab." From course evaluations, students made the following comments about why they disliked the course:

- "The grading, because I did the right work but had bad data and got A LOT of points off.",
- "I felt doing the blue sheet (prelab assignment) and then spending the first 30 minutes or more of the period discussing the lab was too much.",
- "Why is everything solo? Is that real life situation?

Implementing Change

We implemented the changes over a period of three semesters. This allowed time to develop and refine the structure of the course, the laboratory manual and the instructor's manual. This process occurred in several distinct steps and the plan we used to roll out the revised laboratory experiments helped facilitate this development.

Summer 2004 – the Laboratory Manual

The university support from the summer stipend provided an opportunity to spend time rewriting the laboratory manual. Most of the changes were in the structure of the experiments, rather than the content. For example, we kept the "synthesis of copper sulfide" experiment, but made significant changes to the instructions and the laboratory report. By keeping seven of the existing experiments and only adding two new experiments, we simplified the transition from the old laboratory experiments. We started by deciding how the experiments would be structured and what the grading would be based upon. To keep the logistics managable for both the students and the faculty we devised the following structure:

1. Experimental Procedure – Developed by the students working in groups of 3 or 4. A draft was submitted prior to lab for instructor comments, revised for lab based upon these comments, and modified for the final report to reflect the actual experimental procedure the students eventually used.
2. Laboratory Notebook – Kept by each student with the carbon pages submitted at the end of the laboratory period.
3. Laboratory Results – Each group was responsible for preparing a single copy of a spreadsheet and any graphs necessary for each experiment.
4. Discussion – Each student submitted an individual discussion of the group or the class results.

The time spent working out this structure helped us to identify what the most important learning objectives were and to work out a way for the students to meet those objectives. A significant amount of time was spent discussing how to provide the students with the best learning opportunity while keeping the workload requirements for both the students and the faculty reasonable. There are advantages and disadvantages to having students work in groups, and it is important to identify what activities are best suited to each format.

Fall 2004 – Pilot Section

Piloting the new laboratory experiments with an honors section was an extremely valuable experience. Without any prior experience using guided inquiry laboratory experiments, it was difficult to identify how much guidance students needed and which activities would be the most useful learning opportunities. We owe a great deal to these students who were willing to put up with vague and ambiguous instructions and grading criteria. The students in this section were all participants in Widener University's honors program in general education. According to the program description "Traditionally, students who are in the top 10 percent of their graduating classes, have SAT scores of 1200 or higher, and cumulative grade point averages of 3.4 or higher are invited to participate." From this experience we learned how much the students could do in lab, how much guidance they needed to successfully complete the experiment, how much guidance they needed for calculations, and what to expect from discussions.

Based upon the experience with the pilot section we were able to develop the scoring rubrics for each experiment, modify the laboratory manual to provide students the information required to successfully complete the experiment, modify the structure of the course to maximize the student learning opportunities, and reduce the time spent on tasks that did not provide useful learning opportunities.

During the semester, the instructor developed the grading rubric for each experiment as a part of the grading process. Having examples of what the best students did allowed us to revise the laboratory instructions and structure to clarify problems and it also allowed us to identify characteristics that are representative of the best student work. A sample showing part of a grading rubric is shown below in Figure 1.

One of the most important outcomes from the pilot section is that we were able to develop a skills ladder for the most important learning objectives. One of the problems with our previous laboratory structure was that students were expected to be able to do everything at the beginning of the semester. The only way they could do this was by providing them with detailed instructions, which removed the element of concept invention. An important component of our new

Notebook (25 pts)			
Key Data		**Descriptive**	
/20		**/5**	
20	Known & Standards • Mass • Volume • Temperature • Unknown Mass • Volume • Temperature Unknown Number	5	Includes how measurements were made.
18	1 piece missing	4	
15	2 pieces missing	3	Fair
12	3 pieces missing	2	Poor
10	Very poor	1	Very Poor
0	Nothing	0	Nothing

Figure 1. Sample Grading Rubric

laboratory structure is that we don't expect them to be able do everything well until the end of the semester. This allows us to gradually build the student's skills during the semester. The experience we gained from the pilot section allowed us to construct these skills ladders and embed them in the grading rubric. By embedding them in the grading rubric we could clearly communicate our expectations to the students and ensure that students in a course with multiple sections taught by multiple instructors recieved the same experience. For example, the skills ladder for the laboratory notebook includes the following:

1. Key data (ie: mass and volume)
2. Descriptive (ie: how measurements were made and observations)
3. Organized (ie: used tables, information is sequential)

At the beginning of the semester, we were satisfied if the students recorded all of the key data. As the semester progressed, our expectations for more complete descriptions and better organization increased. By the end of the semester, we found that most of the students were maintaining excellent laboratory notebooks. We developed additional ladders for laboratory skills, compilation of results, individual discussions, and group presentations.

Spring 2005 – Off- sequence Semester

In the spring of 2005 we ran the new laboratory experiments with three sections and two instructors. This provided an opportunity to revise the laboratory manual to reflect what we learned from the pilot section and to identify barriers to implementing the laboratory with multiple instructors. A weekly instructors' meeting was held to discuss expectations for the upcoming experiment and to discuss problems that occurred in the previous week. These meetings provided the content for further revisions to the laboratory manual, and allowed us to identify common student mistakes, develop strategies for guiding students, develop timing guides to keep students on task, and to start development of an instructors' guide.

The laboratory manual was revised to include the grading rubrics developed with the pilot section. These were further refined to reflect the expectations of the skills ladder. For example, in the first experiment a student would be expected to record the volume of solution that they used (i.e.: 25 mL) and the instructor would make a comment on the grading rubric that they should also include how the volume was measured (see Figure 2). For the second experiment the student would then be expected to write down the volume and how it is measured (ie: pipet 25 mL) and the instructor would then make a comment that they should describe the solution (ie: pipet 25 mL of clear, colorless solution). Instead of telling the student at the beginning of the semester everything that they have to do, this gradually increases the expectations over the course of the semester. The skills ladder guides the students towards the learning objectives and helps the faculty maintain consistency with grading across multiple sections. This allowed us to develop a much more logical grading structure that both faculty and students clearly understood. Some examples from the skills ladder for student procedures are shown in Figure 3.

The instructor's guide consisted of several sections. The first section detailed the preparations needed at the beginning of the semester, such as scheduling equipment and setting up the classroom management system on our campus intranet. The second part dealt with the connections between the various experiments and the lecture course. This was added to provide the lecture instructors with a timeline so they could schedule discussions of topics to complement the experiments. For example, we wished to coordinate the classroom introduction of percent composition calculations and empirical formulas with the week in which the laboratory students were writing their procedures for the copper sulfide experiment.

The major aim of the instructor's guide was to provide detailed guidance for each experiment as well as for managing a guided inquiry lab. The experiment notes represent a diary of sorts in which the first instructors recorded common student errors and ways to respond to them. The guide was made available in electronic format so that instructors could copy and paste comments into student draft procedures rather than writing the same comments over and over.

Sample Experiment

Experiment 1: Density and Measurements

Goals

1. Students will learn about density, scientific measurements and experimental precision.
2. Students will learn to use graphical information to interpret experimental results.

Objectives

1. Students will be able to use volumetric glassware.
2. Students will be able to use an analytical balance.
3. Students will be able to determine the density of an unknown solution.
4. Students will be able to graph their experimental data.

Introduction

For this experiment your group has been assigned to a forensics laboratory. A grizzly murder has occurred on the Chesapeake Bay near St. Michaels and the police apprehended a suspect last night. The suspect claims that they were fishing near Havre De Grace, Md. The police have collected a 100 mL sample of the water that was in the boat trailer. In addition to the equipment at your lab bench, you have the following equipment available in your laboratory and information from several web sites. Your job is to write up a report for the district attorney explaining your results and clearly presenting proof of your results.

Equipment:

1. Pipet (10 +/- 0.02 mL)
2. Pipet (25 +/- 0.03 mL)
3. Volumetric Flask (100 +/- 0.08 mL)
4. Sauter Electronic Balances (0.001 gram precision)
5. Analytical Balances (0.0001 gram precision)
6. Thermometer
7. Pipet bulb
8. NOTE: Bunsen burners are not available for this experiment

Reagents:

1. Sodium chloride
2. Deionized water
3. Sample of water from suspect's trailer

Figure 2. A Sample Experiment. Continued on next page.

Resources:

1. Maryland Department of Natural Resources: http://mddnr.chesapeakebay.net/
2. Equation of State for Water from Fermi Lab: http://fermi.jhuapl.edu/denscalc.html
3. University Corporation for Atmospheric Research: http://www.windows.ucar.edu/tour/link=/earth/Water/chesapeake.html
4. Chesapeake Bay Program: http://www.chesapeakebay.net/index.cfm
5. National Oceanographic and Atmospheric Administration (NOAA) Coastwatch: http://coastwatch.noaa.gov/seanettles/index.html
6. Units of Concentration. Section 14.1 in Chemistry & Chemical Reactivity

Procedure

1. Volumetric Glassware. Identify and find out how to use different pieces of volumetric glassware. Fill out the glassware template and submit to campus cruiser by the end of lab.
2. Identify how you will determine if the water sample taken from the suspect's boat can connect them with the scene of the crime.
3. Make a graph that shows the information you expect to find from your experiment.
4. Test the procedure using a set of standards. Each student in the group prepares and measures one standard.
5. Use the procedure with your group's sample. Each student runs the sample once.
6. Your procedure should account for the following:
   a. The chain of custody for samples is very important. Your unknown will contain a code number- what do you need to do with this?
   b. The temperature of the sample was not recorded when the sample was obtained. How does the temperature affect the sample? What happens if the temperature changes? Is this related to the properties you are measuring in the lab? Do you need to account for this?
   c. If you have a known volume of water and you add something to it, will this change the volume? What does this mean when you make a standard? Does this affect the concentration of your standard? Does this affect other properties of your standard that you are measuring? How can you design your procedure to account for these changes?

Figure 2. Continued.

Grading

1. Procedure (25 pts, group).
   a. Submit Word document to Campus Cruiser for approval prior to lab
   b. Each student should bring a copy of the procedure to lab.
   c. Each group will submit one printed copy of their final revised procedure with their lab report.
2. Notebook (25 pts, individual).
   a. Before you can begin the experiment you must write, in your own words, the objective of the experiment in your laboratory notebook.
   b. During the experiment record all your data and observations in your laboratory notebook.
   c. Submit carbon copies of laboratory notebook to instructor before leaving lab. It will be returned by the instructor to include with your submission.
3. Results (25 pts, group).
   a. Datasheet. Use template from shared files in campus cruiser to prepare a datasheet. One person in each group should submit this file to the assignments in Campus Cruiser before leaving lab.
   b. Process data and prepare graphs and tables. Use Excel for the calculations and then prepare a table in Word.
   c. One person in each group should submit the Excel file with the calculations via Campus Cruiser.
   d. One printed copy of the Excel spreadsheet should be included with your group submission.
   e. One printed copy of each table or graph should be included with your group submission.
4. Discussion (25 pts, individual). Completed by individual. Submit completed lab report as hardcopy by due date. The report will be a brief summary of your results. The district attorney is very busy, so this needs to be limited to two pages. Use graphs and tables to present the information as efficiently as possible. Your report should include:
   a. Results from your standards
   b. Results from your unknown sample
   c. Supporting references that demonstrate the validity of your method.
   d. Discussion of your results, identify sources of uncertainty and how they could be minimized by changes in procedure.

Figure 2. Continued.

Experiment	Draft Procedure	Final Procedure
1 - Density	Comment on detail (amounts and equipment) and references. Most students will have ignored preparation of standards.	Grade on details and references; comment on safety and waste disposal
2 - Distillation	Comment on safety and waste disposal	Grade on details, references, safety; comment on reference numbers, observations vs conclusions
3 – Synthesis of Copper Sulfide	Comment on reference numbers, observations to make	Look closely at safety information, comment on source, quality, explicit references (not just bibliography)
4 – Synthesis of Alum	All of above	Sources must be properly referenced and of good quality in addition to all of above.

Figure 3. Example Skills Ladder – Procedures

Timetables were developed to give guidelines for each experiment, with information such as "within 1 hour, all students should have their stills running." Lastly, the instructor's guide provided complete versions of the student grading rubrics with the correct answers embedded to assist the grading process.

For both the lab manual and the instructor's guide, we learned quickly that it was wise to begin the next semester's versions at the beginning of the current semester so that changes could be made while they were fresh in our minds. This also reduces the workload at the beginning of the next semester.

Fall 2005 – Rollout

In the Fall of 2005 the new laboratory experiments were rolled out to 120 students in eight sections. These were taught by seven instructors, two with previous experience from the spring, and one adjunct faculty member. A weekly meeting was held to discuss expectations for the upcoming experiment and to discuss problems and issues from the previous week. These discussions were often lively and they were critical to the success of the rollout.

The weekly discussions provided a forum for extensive revisions to the instructors' manual. After these discussions the manual included timing guidelines and information about the most common problems students encountered during the experiment. The discussions also provided an opportunity to exchange strategies for helping students identify and resolve problems that occur during the experiment.

At the end of the semester we held a half day workshop to fine tune the grading rubric. This provided a framework to discuss the goals and objectives for the course and make sure that they were embedded in the structure and grading of the course. We also used this time to assess the new course structure and to make sure we were meeting the objectives we had initially laid out.

Spring 2006 – Maintenance

In the spring 2006 semester, three sections of the course were taught by two instructors. One of the instructors was an adjunct with no previous experience using guided inquiry or teaching these experiments. With the laboratory manual in its fourth revision and the instructors' manual in its second revision, the adjunct was able to step in with a clear understanding of the department's expectations and how to teach the course using the new structure. Based upon student feedback, the revision to the first semester laboratory has been successfully impemented. We now have a general chemistry laboratory program that is designed to meet the departmental goals and objectives for student learning. As we continue to work on program assessment we will continue to evaluate and revise the laboratory course.

Key Steps and Considerations

Based upon our experience, it is clear that we made good choices at several steps while implementing these revisions to our laboratory curriculum. Anyone looking to make significant changes in the structure of the laboratory for a multi-section course should spend time carefully considering these same choices. The important lesson we learned was about the path we found to a successful implementation. Key steps in this were:

- Catalysts for change. We spent several years going to conferences and working on assessment to get the entire department to buy in to a need for change. If we had rushed ahead and tried to implement changes in the structure of the laboratory experiments before establishing a need for change, the process would have been much more confrontational.
- Use of assessment. The steps of assessment are very helpful for laying a groundwork for what change should look like. Time spent discussing the goals and objectives for a course builds consensus and provides a

framework to make future decisions. Clearly establishing the student learning objectives provides a reference to use for deciding what students should do and how instructors should evaluate student performance. Without this reference, faculty will have different ideas about what students should do and how they should be evaluated. For a 100 point experiment we could usually come up with about 200 points to take off for different mistakes students made in their reports. The goals provide a way to identify what counts.

- Phased rollout. By running a pilot section and then expanding to additional sections in an off-sequence semester, we were able to work out a number of small issues before they became unmanageable. This was particularly important in the development of the grading rubrics and skills ladders, which were crucial elements for consistency across a multi-section lab course.
- Skills ladder. This helped us to identify a path for students to achieve the course objectives. For a process-based laboratory experiment, identifying these steps allowed students to develop their abilities as the semester progressed. It also simplified the evaluation of student work because the faculty could focus on specific steps as the semester progressed.
- Meetings. We spent many hours discussing the laboratory experiments, the faculty expectations, the student learning objectives, the skills ladders, problems students had in lab, strategies for guiding students, strategies for resolving group problems, and how to evaluate student performance.
- Instructor's Manual. This document provided a framework for discussion and reviewing what the instructors need to do during lab.
- Revision, Revision, Revision. The laboratory manual and the instructor's manual both went through multiple revisions before they were in final form. As we continue with the assessment of the course and with our program, we will continue to revise this laboratory program.

Conclusions

As stated above, planning was the key to a successful POGIL implementation in a multi-section general chemistry laboratory. By establishing consensus in the department about goals and objectives, and by doing the implementation in steps, we have made a major change in the way we teach laboratories without major roadblocks. We are currently in the process of revising our second semester laboratory course in a similar manner.

Chapter 18

Assessing POGIL Implementations

Renée S. Cole[1] and Christopher F. Bauer[2]

[1]Department of Chemistry and Physics, University of Central Missouri, Warrensburg, MO 64093
[2]Department of Chemistry, University of New Hampshire, Durham, NH 03824

How will an instructor know whether students are benefiting from POGIL? How can an instructor get the most out of the initial and subsequent implementations of POGIL? Assessment must include a feedback loop that allows the instructor to identify the strengths and areas to focus on for improvement. This assessment information can come from self-analyses, student assessments, and peer assessment from other instructors. The different assessments can be used for three different levels of analysis: a specific activity (a guided inquiry worksheet, a particular demonstration, a lab experiment), a general component of the course (lab, group work, lecture, the text, etc.), and the course in general.

End-of-semester surveys completed by students for the purpose of evaluating a course and its instructor are part of the normal routine of higher education. The institution seeks information of this type to respond to calls for accountability, to provide input for promotion and tenure decisions, and to provide instructors feedback about their instruction and course structure. These surveys attempt to answer questions that are summative in nature. They are analogous to the final grade received by students – an indication of the overall performance for a semester of effort. Much effort has gone into studying the nature of end-of-semester evaluations and their interpretation (*1-8*).

We are not going to explore that literature here because end-of-semester evaluations provide information that is too broad and too late to help with a curriculum implementation. When instructors decide to implement POGIL, they will have some sense nearly immediately as to how well students are adapting to the new expectations, and as to how well they are able to manage the new expectations they have of themselves. This feedback will inform the planning of the next day's activities. In Le Chatelier terms, introducing POGIL has put a stress on the instructional system, the system is now in flux, and it will eventually establish a new equilibrium. Unlike molecular systems, which respond blindly to the stresses, the human classroom needs to be managed. Thoughtful management requires good data. Good data will come from frequent assessment of where students are conceptually and affectively, and of where the instructor is in his/her implementation.

Studies of curriculum implementation contend that one of the most important features of a successful implementation is to monitor the levels of concern of those involved and respond to address those concerns (*9-14*). The interventions are nuanced to take account of how much experience the users have and how they fit within the social network. This is particularly important for a first implementation because, despite pre-course planning, an instructor will likely be living day-to-day in terms of class preparation.

By having participants feel that they have some ownership or choice in the implementation process, the level of cooperation and trust is higher. It becomes something they are doing together with the instructor, instead of something that the instructor is doing *to* them.

Instructors should also realize that they are not alone. Many chemical educators have engaged in implementing POGIL (*15*) and have wisdom to share about implementation and assessment. They can also be a source of emotional support should the going be rough with students or colleagues.

Assessment must include a feedback loop that allows the instructor to identify the strengths and areas on which to focus for improvement. This assessment information can come from self-analyses, student assessments, and peer assessment from other instructors. The different assessments can be used for three different levels of analysis: a specific activity (a guided inquiry worksheet, a particular demonstration, a lab experiment), a general component of the course (lab, group work, lecture, the text, etc.), and the course in general.

Assessment Tools

Successful assessment of a POGIL implementation requires familiarity with appropriate tools to provide feedback. The focus of the assessment (activities, course components, or the course in general) will determine which of these are most appropriate to collect information. Who conducts the assessment may also vary among self, students, and peers. Some tools can be utilized by all three constituents; some are more suited to a particular group. We will highlight SII (Strength, Improvement, Insight) analyses, surveys, classroom observations, student interviews, course journals or notebooks and annual reports. Our focus on classroom assessment is not novel (*16*), but current classroom practices of many instructors suggests that detailed documentation and reflection on implementation of new instructional ideas is not yet routine.

SII Analysis

SII stands for Strengths, areas for Improvement, and Insights (*17*) . This is an effective tool for analyzing a product or process in an explicit way and collecting useful feedback. The repeated use of this strategy also improves the assessment skills of those using the instrument and is not as threatening as many "evaluation" techniques. The person with the process or product to be assessed should define the objectives and outcomes of what will be assessed. This serves as a reference point for the person doing the assessment to focus feedback on these areas of concern. The assessor first records the strengths of the product/process and indicates why they are strengths for achieving the designated outcomes. The next step is to identify areas of improvement for the product/process as well as how those improvements could be made and why they would be improvements. The last step is to list any insights that were discovered as a result of the assessment process. The completion of an SII analysis often provides as much information to the assessor as to the person having the assessment done. When someone is inexperienced in the role of assessor, it is helpful to conduct an SII on the feedback. This is particularly true if the assessor interprets the role as "critic" and does not want to say anything "negative". Conducting an SII on the feedback defines what desirable feedback is for both the assessor and the person assessed, establishing trust in the process and "permission" to provide critical comments. It also helps refine the information desired.

Having students complete SII analyses may require some development of their skills. Students generally do not have much experience in assessment and may find the process difficult. There are advantages in having students complete some preliminary assessments to become familiar with the process. An additional benefit is that students learn to assess their own efforts and products more effectively. One way to do this is for students first to complete a peer assessment of a product (exam question, essay, draft of a project, lab report, etc.) using SII. Then students complete an SII of the feedback to share what the strengths of good feedback are and what they thought would have improved the feedback they received. Assembling a list of student perceptions of good feedback provides students ownership in the process since they generate the list instead of having it dictated to them. Receiving input on something they have created also communicates the value and importance of feedback in the production of quality materials.

Surveys

There are many different types of surveys, although the most common in courses are those with a pre-determined scale (Likert-type surveys). Surveys can be used at many different points in the course depending on one's goals: for instance, mid-term, before or after an exam, after completing a particular activity, or at the end of the semester. An instructor can also combine surveys with more open-ended assessments. Some examples of surveys that can be used for course assessment can be found on the POGIL website (*18*). An instructor can also generate similar surveys on the SALG website (*19*). If a goal is to generate publishable data, survey validity and reliability characteristics should be established (*20-23*).

Classroom Observations

Classroom observations can provide many insights into the learning process (*24-26*) . Observations are most effective if focused on specific target behaviors. An instructor can make notes while class is occurring, although it is difficult to maintain distance from the instructional interactions with students in order to monitor broader components. Videotaping allows an instructor to be an observer after-the-fact. However, there may be difficulties in getting an accurate observation of an active class such as POGIL from a static viewpoint.

Observations by a peer provide both the observer and the observed important information about the teaching and learning process. An instructor can have a single individual observe the class or can have a group of individuals observe. With an individual observer, it is most effective to first complete a form that outlines the learning outcomes for the class period being observed. These outcomes should include the content knowledge, skills, and attitudes that the instructor desires to develop during the observation time. Observers then complete an SII analysis of how well these goals were accomplished based on their observations. They can point out things done well, missed opportunities, and roadblocks. If there are multiple observers, it is frequently useful to have each observer focus on a different aspect of the class. For example, one observer may be focusing on the class as a whole, while another focuses on a single group or one observer can focus on development of concepts and another on the development of skills.

Student Interviews

Student interviews can provide additional insights into the impact of different aspects of a course. These can be conducted with individual students or with small groups of students. The interviews can be performed either by the instructor or an external observer. The format can be informal (questions asked during or after class) or more formal (questions planned beforehand and interview times arranged). Conversations with students often elicit different information from what can be obtained in written comments or surveys. A traditional way to interview students is to have a set of questions to ask students about a particular item or process (*27-29*). Another approach is to have students complete a task and have them think aloud during the process (*30-32*) or to stop students at certain points and ask them why they did or didn't do something. This process would be useful if an instructor wanted to obtain information on the impact of POGIL on students' problem solving methods or critical thinking.

Journals

It is useful for the instructor to keep a notebook or journal to document course materials and activities as they occur. This provides an organizational tool to keep materials such as student handouts and instructions, and it also provides a record of how a topic was taught in the most recent instance of the course. It is a good place to record notes of what worked well and where improvements are needed, such as the records of SII analyses. This provides a springboard for course improvements the next time the course is taught.

Annual Report

Documenting achievements and identifying areas for growth on a regular basis is a useful strategy to both reflect on the accomplishments and progress made over a year and provide the motivation to identify the areas on which to focus for the coming year. Reflection puts into perspective how much has been accomplished. Often curriculum reform seems like a long, slow process with very little progress made. The act of looking back should provide a sense of satisfaction and encouragement that more progress is being made despite how it may appear on a day-to-day basis. This annual documentation is also useful for establishing effectiveness as a teacher for promotion and tenure purposes.

Assessing Activities

Assessment may be conducted at the level of individual activities, components within a course, or the course as a whole. We will consider each of these in turn.

An instructor may be writing his/her own POGIL materials, which could include classroom activities or laboratory activities. The process of developing materials requires several iterations to achieve an optimal product. A structured assessment of activities will not only provide important information for the improvement of a particular item, but will also provide insight into the process to improve the development of new items.

Self (Activity Assessment)

An initial self analysis will focus on whether the intended outcomes of the activity have been addressed and whether the activity meets the general guidelines for POGIL activities. The analysis can be done more consistently using the two rubrics designed to make sure all desired components are present and sufficiently developed. These rubrics can be found on the POGIL website (*18*). The next step of analysis comes when the activity is used with students for the first time. It is beneficial to obtain explicit data from the students regarding the activity (outlined below), but the instructor will also want to make informal observations and notes. When the activity is initially used with students (and after significant modifications have been made), one should pay special attention to areas where students seem to get stuck, are confused, or are particularly frustrated. These indicate areas that may need more information, additional guiding questions, or just small changes in language to clarify what students should focus upon. Activities may be designed for implementation in the laboratory, lecture room, recitation, or for take-home practice. It may be desirable to try the activities in a variety of implementation formats to observe how this affects the outcomes.

Student (Activity Assessment)

Students have a vested interest in well-written materials and can provide valuable insight into how well an activity is meeting the intended outcomes. A powerful tool is to have students conduct an SII analysis of the exercise. There are two ways to do this. The instructor can identify the intended outcomes (including knowledge, skills, and affective responses) and have students identify the strengths, areas for improvement, and their insights into the topic or process. Another method is to have students themselves identify what they perceive as the intended outcomes and then complete the SII components. Having students articulate what they believe they learned by completing the activity, when contrasted with the instructor's intended goals, provides valuable feedback for modifying the activity to better address those outcomes and provide students with a clearer orientation as to the goals of the activity.

The instructor may also want to construct a quick survey for students to complete after the introduction of a newly written or edited activity. This would ask questions related to the degree to which the activity met the goals intended by the instructor and could include questions related to developing skills and attitudes as well as about structure and format. A Likert scale survey provides a quick snapshot of student perception, but will not provide as much information as an open-ended questionnaire (such as the SII technique outlined above).

Interviewing students is another good way to obtain data about an activity. This is time intensive and not feasible to use with every new activity. Student interviews can be conducted individually or with focus groups, for example, drawing from only the group managers for that activity. These interviews can be particularly valuable early in the writing process to gain insights into student needs and approaches to learning. An instructor may also want to interview students informally during the classroom implementation. This is easily done by asking students questions related to activity design while they are completing the activity, particularly when student behaviors are observed that were not expected or anticipated. Having students describe their reasons will provide the instructor with additional insights into how students process information and react to the activities being developed.

Peer (Activity Assessment)

Colleagues are an important asset in developing new materials. They could be at one's own institution or they might be experienced POGIL users at other institutions. Being willing to share the new materials with them upon completion may be a helpful incentive to enlist colleagues in this endeavor.

The first step is to identify the key learning outcomes for the activity. These should include the knowledge, skills, and attitudes the instructor is trying to develop through the implementation of the activity. The instructor may also need to supply some basic information regarding the activity such as the course, pre-requisite knowledge, and target student population to enable the assessor to provide more detailed and useful feedback. The next step is to have the peer conduct an SII analysis of the activity. The instructor may also want to provide the assessor with a general rubric (*18*) for the format of POGIL activities to have them assess how well those components have been met in addition to conducting an SII analysis of how well the activity meets the intended learning outcomes. For example, how is cooperative group structure set and managed by the activity, to what extent does the instructional sequence exemplify an appropriate learning-cycle organization, or does the content agenda for learning allow student entry and build to important ideas? As mentioned in the description of SII analysis, an instructor may want to complete an SII of the initial peer feedback to develop the peer observer's ability to provide detailed, useful feedback on the activities.

Assessing the Course

Making a commitment to implement POGIL will require substantial thought and planning. The instructor (and his/her colleagues, department chair, and dean) will be interested in knowing answers to questions such as "How have the students changed?" "What's different about their learning?" or "Is the effort worth it?" Answering questions such as these in a strong way will facilitate a supportive environment to continue using POGIL or to encourage colleagues to begin using POGIL. Consequently, the time to think about course assessment is not in the middle of implementation, but beforehand in order to establish a baseline on which to draw comparisons. Whatever types of analyses or assessments are used to assess POGIL once it is in place can be applied to the course in its continuing form. Another reason to plan ahead is that, at most institutions, the course experiment can't be repeated more frequently than every four months (every semester). Thus, the process of review, revision, and retrial may play out over a period of years. Poor planning or hastiness may lead to compromised or aborted implementations, weak assessments, and wasted time, and may lead to a poor perception of POGIL among the student population.

A few guidelines have been found by other POGIL implementers to be helpful in developing course goals and to delineate areas for improvements. Generally, course goals should be:

- have measurable outcomes.
- be written using positive language.
- be linked to classroom activities and contexts.
- increase the degree of ownership students feel regarding the course.
- motivate change or growth in students or instructor.

We will now explore how the course as a whole might be assessed by oneself, by students, and by faculty peers.

Self (Course Assessment)

The person introducing POGIL will be interested in how well implementation is proceeding and whether desired outcomes are being achieved.

It is important to save time for this process and to document one's thinking in a journal or laboratory notebook as a record of progress. With a new implementation, this type of activity will be a daily event. One may want to share the plan with another individual. This could be a colleague, department chair, or peer coach. Documenting this process is also a useful component of a promotion/tenure dossier as evidence of teaching effectiveness.

Student (Course Assessment)

Obtaining student feedback before the end of the semester is not a new suggestion (*16*). Changes made based on end-of-semester information can only benefit the next set of students. Students involved in the POGIL implementation will be more invested if they are able to provide early feedback, if they hear a summary of that feedback, and if they see reasonable adjustments to or engage in substantive discusion of course requirements and expectations. Anonymity has pros and cons (*33*) and may be implemented as desired. Participation can be encouraged by awarding points. In larger classes, feedback may be solicited from all but can be sampled to reduce processing effort. At a minimum, students should have two opportunities to provide feedback: at the middle and end of the semester.

The approaches an instructor might use include the SALG, an SII inventory of the course as a whole, and/or an SII inventory by the student of themselves as a learner in the course.

Peer (Course Assessment)

As scientists, we seek out the expertise and opinions of peers to help us reflect on our own ideas, experiments, and evidence. We do so in order to improve our understanding of those ideas and improve our next round of experiments. Why should a curriculum implementation be any different? An instructor could ask a colleague from the department, from another discipline, or from another institution (e.g. POGIL regional or national network) to assist in being another pair of eyes and providing another perspective. Specifically, one might be interested in having a colleague observe the class at work.

Direct the colleague to observe two to three specific features of an instructional session or class, such as student behaviors, diversity of student engagement, cognitive level of questions or comments, or evidence regarding understanding of chemical ideas. The observer should make a written record of these observations. The value of this report would be improved if the observer did an SII analysis of the instructional event. Additional insights may be obtained if the colleague is willing to interview a few students as a group or individually. This is a win-win-win situation. The instructor benefits from the insights from the observations, the students benefit from knowing that their input is valued, and the observer benefits from participating in a reflective process, which she might then apply to her own teaching. One can complete the assessment cycle by having the instructor provide an SII analysis of the peer feedback, which establishes a learning dialog for both the observer and observee (*34*). This process should be repeated two or three times during the semester to provide feedback on the progression of the implementation and note improvements resulting from earlier assessments. The observer could also be a graduate or undergraduate student, but their vision will not be as experienced as that of another colleague and their status may not free them to comment candidly about what was seen.

Assessing Course Components

If the instructor is focusing on implementing POGIL in a particular component of the course (i.e., laboratory, lecture, recitation), it will be desirable to assess the impact on that component of the course in addition to the course in general. This can be particularly helpful in optimizing the implementation of POGIL instruction.

Self (Component Assessment)

The initial process will be very similar to that for course assessment. The instructor will first want to identify the strengths and areas for improvement for the course component being changed. The instructor will want to consider the desired student learning outcomes to be developed by that aspect of the course and make sure that the changes being made will promote the development of the identified outcomes. Keeping a journal and comparing the outcomes during the implementation to previous semesters is an informal way to gather information concerning the impact of the implementation. The instructor can also assess student progress in developing the desired outcomes through exams, quizzes, lab practicals, etc. If data has been collected in previous semesters or on previous assignments, this will facilitate comparisons.

Student (Component Assessment)

Students can complete SII analyses of the course component under consideration. A survey could be developed as a stand-alone instrument to collect information about student perceptions of the course component or it can be a focused part of a general course survey. The questions should focus on the impact of the changes made in the course component on helping students achieve the desired outcomes.

Peer (Component Assessment)

The process for obtaining peer assessment is identical to that for the course in general, but the focus should be on a single aspect of the course rather than the course in its entirety. The key to obtaining effective feedback is to provide the colleague with desired outcomes for that aspect of the course.

Summary

This brings us back to the original questions. How does an instructor know if students are benefiting from the implementation of POGIL? How can an instructor get the most out of initial and subsequent implementations of POGIL? The feedback loop provided by these assessments allows the instructor to identify the strengths and areas to focus on for improvement. It is an ongoing process practiced by engaged, productive faculty who want to improve student learning and experience personal growth as an instructor. Even "expert" POGIL users continually assess and improve their implementation of POGIL in their classroom instruction and look for opportunities to exchange ideas and information with other instructors. This cycle of assessment and change is characteristic of the POGIL philosophy that learning is an interactive process of refining one's understanding and developing one's skills.

References

1. Cohen, P. A. *J. Dent. Educ.* **1991**, *55 (2)*, 145-150.
2. Marsh, H. W.; Bailey, M. *Multidimensional Students' Evaluations of Teaching Effectiveness: A Profile Analysis*; New South Wales, Australia, July 24, 1991. (Educational Research Information Center: ED350310)
3. Hansen, W. L.; Kelley, A. C., *J. Econ. Educ.* **1973**, *5 (1)*, 10-21.
4. Drews, D. R.; Burroughs, W.J.; Nokovich, D. *Teach. Psychol.* **1987**, *14 (1)*, 23-25.
5. Marsh, H. W.; Overall, J.V.; Kesler, S.P. *Am. Ed. Res. J.* **1979**, *16 (1)*, 57-69.
6. Williams, R. L. *Course Evaluations: A Strategy for Improving Instruction*; University of Tennessee: 2001. (Educational Research Information Center: ED449759)

7. Freidman-Erickson, S.; Waller, M. P. Enhancing the Usefulness of Students' Evaluations of Instruction, *Annual Meeting of the APS Institute on the Teaching of Psychology*, New York, NY, June 1995. (Educational Research Information Center: ED386632)
8. Aleamoni, L. M., *J. Pers. Eval. Educ.* **1999**, *13 (2)*, 153-166.
9. Alper, L.; Fendel, D.; Fraser, S.; Resek, D., *Am. J. Educ.* **1997**, *106 (1)*, 148-178.
10. Churchman, D. A.; Hellweg, S. A. Measuring Student Needs for University Programs and Selecting Appropriate Curricular Designs, *Society for College and University Planning Conference*, Kansas City, MO, July 1979.
11. Tobin, K.; Dawson, G. *Educ. Technol. Res. Dev.* **1992**, *40 (1)*, 81-92.
12. Kallery, M.; Psillos, D. *Eur. Early Childhood Educ. Res. J.* **2002**, *10 (3)*, 49-61.
13. Schneider, R. M.; Krajcik, J. *The Role of Educative Curriculum Materials in Reforming Science Education*; University of Michigan: 1999.
14. Shavelson, R. J.; Borko, H. *Educ. Horiz.* **1979**, *57 (4)*, 183-189.
15. *POGIL Project*, URL http://www.pogil.org/resources/schools.php. Last accessed October, 2007.
16. Angelo, T. A.; Cross, K. P. *Classroom Assessment Techniques: A Handbook for College Teachers*, 2nd ed.; Jossey-Bass: San Francisco, CA, 1993.
17. Apple, D. K.; Beyerlein, S. W. *Faculty Guidebook: A comprehensive tool for improving faculty performance*; Pacific Crest: Lisle, IL, 2006.
18. *POGIL Project*, URL http://www.pogil.org. Last accessed October, 2007.
19. Seymour, E. *Student Assessment of Learning Gains*, URL http://www.wcer.wisc.edu/salgains/instructor. Last accesssed October, 2007.
20. Herman, J. L. *What Do the Test Scores Really Mean? Critical Issues in Test Design*; 1986. (Educational Research Information Center: ED279682)
21. Zwarts, M. A. *Eval. Educ. Int. Progr.* **1982**, *5 (2)*, 119-139.
22. Russ-Eft, D. F. *Validity and Reliability in Survey Research.* ; Technical Report No. 15.; National Center for Education Statistics: Palo Alto, CA, 1980; p 143. (Educational Research Information Center: ED279726)
23. Litwin, M. S. *How to Measure Survey Reliability and Validity*; Sage Publications, Inc.: Thousand Oaks, CA, 1995.
24. Jacobs, J. H. *J. Res. Sci. Teach.* **1973**, *10 (3)*, 213-220.
25. Peters, R. O. *Classroom Observation Criteria and Techniques*; 1976. (Educational Research Information Center: ED121790)
26. Ivany, J. W. G.; Neujahr, J. L. *Sci. Teach.* **1970**, *37 (2)*, 31-34.
27. Dana, N. F.; Dana, T. M.; Kelsay, K. L.; Thomas, D.; Tippins, D. J. *Qualitative Research in Education*, Athens, GA, 1992. (Educational Research Information Center: ED349308)
28. Griffee, D. T. *J. Dev. Educ.* **2005**, *28 (3)*, 36-37.

29. Patton, M. Q. *Qualitative Evaluation and Research Methods*; SAGE Publications: Newbury Park, CA, 1990.
30. Bowen, C. W. *J. Chem. Educ.* **1994**, *71*, 184.
31. Genest, M.; Turk, D. In *Cognitive Assessment*, Merluzzi, T.; Glass, L.; Genest, M., Eds.; The Guilford Press: New York, 1981; pp 233-269.
32. Jeon, K.; Huffman, D.; Noh, T. *J. Chem. Educ.* **2005**, *82*, 1558.
33. Golembiewski, R. T.; Billingsley, K. R. *Group Organ. Stud.* **1976**, *1 (4)*, 448-453.
34. Katz, J.; Henry, M. *Turning Professors Into Teachers: A New Approach to Faculty Development and Student Learning*; Macmillan Publishing Co.: New York, 1988.

Chapter 19

A Multi-Institutional Assessment of the Use of POGIL in Organic Chemistry

Andrei Straumanis and Emily A. Simons

Department of Chemistry, College of Charleston, Charleston, SC 29424

Assessments of POGIL instruction in organic chemistry at seven institutions indicate that POGIL reduces attrition without lowering standards, improves student learning, and promotes the development of key process skills such as critical thinking, teamwork and self-assessment. In both POGIL and lecture sections, common exams including the ACS Organic Exam were used as a basis for comparing student learning, and the Student Assessment of Learning Gains (SALG) survey was used as a measure of growth in process skills. Over 1000 surveys collected from seven institutions indicate that fewer than 8% of POGIL students are negative about the method. The similarities of the findings across the different settings provide general evidence for the effectiveness of POGIL.

Organic Chemistry is rarely an elective and has earned a reputation for being a graveyard for many budding health, biological, or chemical science careers. The subject is highly cumulative, making a conceptual approach to the material very desirable and a comprehensive final exam a good measure of student learning. Like all science courses, it is an excellent medium for modeling and improving scientific inquiry skills. Finally, organic chemistry serves as a key gateway in the pursuit of many careers that are traditionally associated with economic prosperity and influence in our society. In other words, the demographics of today's successful organic chemistry students largely determine the demographics of the next generation of physicians and scientific role models. All of these factors make Organic Chemistry an important target for educational reform, and a rugged testing ground for Process Oriented Guided Inquiry Learning (POGIL).

Traditional organic chemistry courses taught using lecture have notoriously high attrition rates. The average D, F, or W (withdrawal) rate from lecture-taught first semester organic courses reported in this chapter is over 30%. This rate is more than double for students who self-identify as Black, Hispanic, or other. A large number of factors likely contribute to these high attrition rates, but the literature suggests that paramount among these are the feelings of isolation and disorientation to which many minority students are particularly susceptible due to the racial separations imbedded in our society (*1*).

POGIL's central claim is that it helps students simultaneously develop content knowledge and key process skills. This chapter addresses both parts of this claim. The first claim, that it allows students to develop content knowledge, is investigated using common final exam data from four institutions (Institutions A-D). The second claim, that it helps students develop key process skills, is investigated using student perception data collected from four institutions (Institutions D-G) using the Student Assessment of Learning Gains (SALG) survey. Profiles of Institutions A-G are shown in Table I.

Assessment of Student Learning Using Common Exam Results

Table II shows common exam results for each of Institutions A-D. At each institution data was collected in both POGIL and lecture sections of organic chemistry. POGIL sections were taught using POGIL exclusively, and lecture sections were taught using lecture exclusively. Although different exams were administered at each institution, the same grading scale was used for each pair of pie graphs, and Pearson chi-square (χ^2) analyses comparing the frequencies of grades of [A], [B], [C] or [D, F, W] for POGIL versus lecture are reported for each pair. *P*-values < .01 show a statistically significant difference at the 99% confidence level between distributions associated with each method. Any student who withdrew from the course prior to the final exam is represented in the W portion of the pie.

There are a number of differences in the study designs and data collection at the four different institutions shown in Table II: at Institution A the lecture and POGIL sections met at the same time and the comparison is based on both the final exam and the hour exams, all of which were common to both sections and graded collaboratively by both instructors. At Institution B students from both a POGIL and a lecture Organic 1 course were enrolled in the same Organic 2 course taught by a third instructor using lecture. Performance in this Organic 2 course provided a basis for comparison between the two different Organic 1

Table I. Profiles of study institutions based on Carnegie Classifications (2)

A	*B*	*C*	*D*	*E*	*F*	*G*
Private	Private	Public	Public	Private	Private	Public
Small	Small	Large	Large	Small	Small	Medium
Highly residential	Highly residential	Non-residential	Primarily residential	Highly residential	Highly residential	Non-residential
Lib. Arts College	Lib. Arts College	Research Univ.	A&S + Profession	A&S + Profession	A&S + Profession	A&S + Profession
More selective	Highly selective	Less selective	More selective	More selective	More selective	Less Selective
N. East	Mid-West	West	South	South	N. East	Mid-West

sections at Institution B. At Institution C students in one POGIL and one lecture section of Organic 1 were given the same multiple choice final exam which was prepared by the lecture section instructor.

At Institution D all sections of Organic 1 and Organic 2 were given the same final exam over a two year period. This provides a means of directly comparing 184 final exam scores from four POGIL sections with 528 exam scores from twelve lecture sections. The Organic 1 and 2 final exams taken by all these students were also given in two prior years providing an additional ten sections of Organic 1 and six sections of Organic 2 for comparison. These sections were of comparable size, averaging 45 students. Lecture sections were taught by one of five different professors, while all POGIL sections were taught by the same professor.

The Organic 1 final exam at Institution D was a multiple choice exam compiled primarily by one of the lecture section instructors using language and material from end-of-chapter textbook questions. The Organic 2 final exam at Institution D was the 2002 American Chemical Society Organic Exam, a comprehensive, standardized, multiple-choice exam. Though the key variable of instructor is confounded with POGIL, the large amount of data collected from lecture sections at Institution D allows us to make some estimate of the impact of this variable on exam performance (see Discussion of Exam Data).

It should be noted that because of the registration patterns of students at Institution D during the years of concurrent POGIL and lecture instruction there was a significant subset of students (N = 35) who had lecture for Organic 1 and POGIL for Organic 2; however, this subset is not reported separately because no significant differences were found in Organic 2 exam scores between these students and those who had POGIL for a full year. The cohort of students who had POGIL followed by lecture was too small (N = 5) to be reported due to human subjects research restrictions.

Table II. Common Exam Grade Distributions

Key	A (black)	B (dark grey)	C (light)	D, F or W (white)	
Inst. Course	Institution A Organic 1	Institution B Organic 2	Institution C Organic 1	Institution D Organic 1	Institution D Organic 2
POGIL	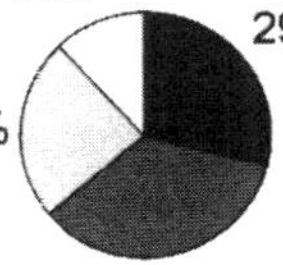 N = 16	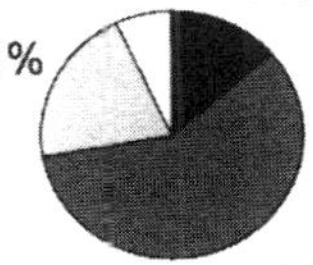 N = 15	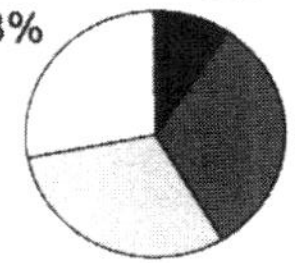 N = 75	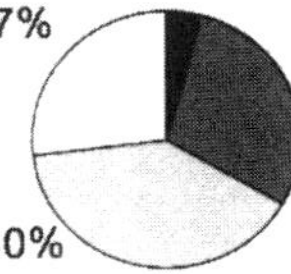 N = 93 (2 sections)	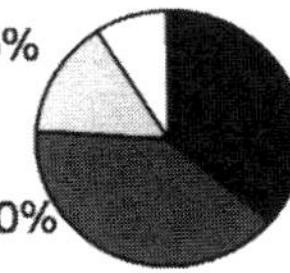 N = 91 (2 sections)
Lecture	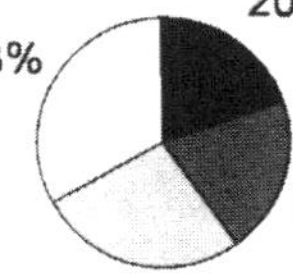 N = 20	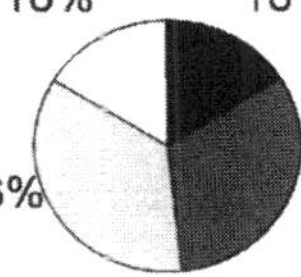 N = 26	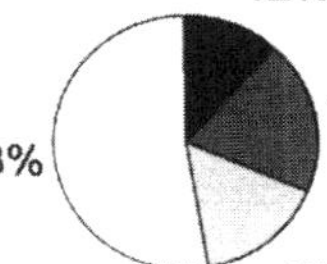 N = 109	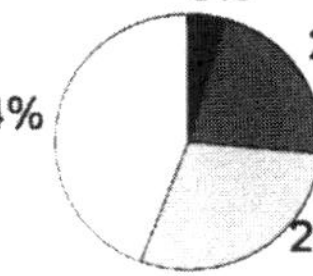 N = 793 (18 sections)	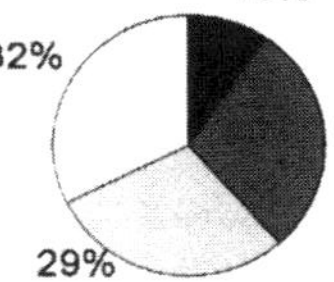 N = 388 (11 sections)
Sig.	χ^2 (3, N=36) = 2.17, p = 0.22	χ^2 (3, N=41) = 4.96, p = 0.24	χ^2 (3, N=184) = 29.30, p<0.01	χ^2 (3, N=886) = 10.76, p=0.013	χ^2 (3, N=479) = 57.70, p<0.01

Discussion of Exam Data

Taken together, the data from Institutions A-D suggest that POGIL has a significant positive impact on student performance in organic chemistry. This effect is strongest among those students who would be predicted to receive a D, F or withdraw from the course, though there is also evidence that POGIL improves exam performance at the top of the grading scale, especially in Organic 2.

The similarity of findings at the four different institutions is suggestive of a general trend; however, it should be noted that only the results from Institutions C and D have a large enough number of cases to achieve statistical significance. In addition, there are two other explanations for the observed results that must be explored as alternates to the conclusion that POGIL has a positive impact on student exam performance. The first alternate explanation is that stronger students self-select into POGIL sections; the second possibility is that the quality of the POGIL instructors is higher than the quality of the lecture instructors, independent of differences between POGIL and lecture. The second of these was not directly controlled, but there is evidence that differences between POGIL and lecture instructors cannot fully account for the observed differences in student performance by teaching method.

If the observed differences between methods result because POGIL sections attract stronger students, we might expect to see evidence of this in other measures of student aptitude. To search for such effects, five independent measures of student aptitude were analyzed by teaching method at Institution D: Math SAT, Verbal SAT, General Chemistry 1 grade, General Chemistry 2 grade, and other course GPA in the semester of study. An analysis of variance (ANOVA) using these variables showed no differences by method at the 95% confidence level.

The impact of instructor on student exam performance is much harder to measure in this case because the variable instructor is confounded with method for the POGIL sections. To deal with this we calculated the effect size of method on exam scores and compared this to the average effect size of instructor on exam scores. The latter was calculated using the large amount of data collected from lecture sections at Institution D. The results described in Table III indicate that calculated differences between instructors in Organic 2 are not large enough to account for the observed differences by teaching method; however, in Organic 1 the calculated differences between instructors *are* large enough to account for differences otherwise attributable to teaching method.

To calculate effect sizes while taking into consideration the large number of withdrawals from lecture sections we used an analysis of rank called a Kruskal-Wallis test. Mean rank is used instead of mean score because the former takes into account both attrition prior to the final exam (W%) and score. It is calculated by ranking all students, then taking the mean of student ranks for a given cohort. The Kruskal-Wallis test showed significant differences in mean rank when analyzed by teaching method or instructor. Teaching method: Organic 1, χ^2 (1, N = 886) = 10.25, $p < .01$; Organic 2, χ^2 (1, N = 479) = 58.20, $p < .01$; Instructor: Organic 1, χ^2 (5, N = 886) = 31.66, $p < .01$; Organic 2, χ^2 (4, N = 479) = 65.59, $p < .01$.

Table III shows a tabulation of Kruskal-Wallis test results including effect sizes by method and instructor. The small effect of instructor reported in Table III is enough to account for the small differences in mean rank between POGIL and lecture in Organic 1, but not large enough to account for the moderate to large differences in mean rank between POGIL and lecture in Organic 2.

To further investigate the hypothesis that the variable instructor is largely responsible for differences attributed to the variable teaching method, we examined institutional end-of-semester student opinion surveys and found that the instructor of the POGIL sections (Q) was not the highest ranked instructor based on this measure. (Instructor T was highest.) This finding is corroborated by the fact that the lecture section instructors appear in a different order in the Organic 1 and Organic 2 parts of Table III. If instructor were the most important variable for determining mean rank we would expect these orderings to be similar. The fact that Instructor Q tops both lists is therefore more likely due to the variable *teaching method* than the variable *instructor*.

Note that Kruskal-Wallis tests by method using only the two years of *concurrent* lecture and POGIL data were also carried out. The effect sizes of teaching method on rank for this subset are slightly larger than those calculated using all available lecture data, since lecture performance was slightly lower during the two years of concurrent lecture and POGIL instruction (Organic 1: η^2 = .023, Organic 2: η^2 = .14). We have chosen to use all four years of lecture data in our comparisons since these result in more conservative analyses.

Assessment of Growth in Process Skills Using Results from a Student Perception Survey (SALG)

Exam performance is not the only measure of the success of a teaching method. The philosophy of Process Oriented Guided Inquiry Learning is, in fact, that a focus on key process skills such as critical thinking, teamwork, and self-assessment will yield improvement in content knowledge, and moreover, that such skills are valuable unto themselves since they form the foundation of scientific inquiry and collaborative work in general.

The following data show significantly higher growth in process skills among POGIL Organic 2 students in comparison to lecture Organic 2 students. Additionally, they suggest that POGIL students take better advantage of course elements and like their course better than lecture students. Organic 1 data were collected and show similar trends, but are not reported here.

Table III. Effect Sizes of Instructor and Method on Mean Rank at Institution D

	Method/ Instructor	N	W %	Mean Rank	Mean	SD	Variable	Effect Size (η^2)
Organic 1	POGIL	93	6 %	523	62.0	17.5	Method	.012 (small)
	Lecture	739	27 %	434	62.0	17.2		
	Instr. Q	93	6 %	523	62.0	17.5	Instructor	.026 (small)
	Instr. R	143	20 %	509	67.4	16.7		
	Instr. S	43	23 %	489	67.0	18.3		
	Instr. T	142	12 %	440	56.2	18.6		
	Instr. U	205	33 %	411	62.3	15.9		
	Instr. V	260	33 %	399	61.3	18.6		
Organic 2	POGIL	91	5 %	339	80.2	11.8	Method	.12 (large)
	Lecture	388	16 %	217	68.5	13.4		
	Instr. Q	91	5 %	339	80.2	11.8	Instructor	.021 (small)
	Instr. V	61	13 %	236	69.8	13.8		
	Instr. T	130	20 %	233	72.5	11.8		
	Instr. R	65	17 %	216	68.6	12.7		
	Instr. U	132	14 %	192	64.2	13.8		

The Student Assessment of Learning Gains (SALG) survey was used to indirectly measure growth in process skills based on the assumption that student perceptions of growth are indicative of actual growth. The survey was administered at the end of the second semester of organic chemistry to 218 POGIL students at four institutions (D, E, F and G) and 188 lecture students at two of these institutions (D and E), all described in Table I. Table IV shows responses by institution, instructor and section. The response rate was above 90% for every section except at Institution F, where no extra credit was offered for completion of the survey.

The SALG is an anonymous, online survey that asks students to rate on a 5 point Likert scale certain aspects of their own learning experience in a given course. It was designed as an antidote to end-of-semester student opinion surveys that focus on the instructor, and are regarded by some as popularity polls, not useful course assessments (*3*).

There are two separate sections to the SALG survey. Items for each part are listed in Table V. Note that each item mean reported in Table V is the average of 9 section averages (for the POGIL column) or 6 section averages (for the lecture column). Each item mean is *not* the mean of all student responses to that item since such a reporting would bias the mean toward the results of large sections.

Part A SALG questions ask students to rate the contribution of various course elements (e.g. instructor presentations, feedback, the text, etc.) to their learning. Part B questions ask students to rate their growth with respect to

Table IV. Response to SALG survey (Organic 2)

<table>
<tr><th>Method</th><th>Institution</th><th>Instructor</th><th># Responses</th></tr>
<tr><td rowspan="9">POGIL</td><td rowspan="2">D</td><td rowspan="2">Q</td><td>43</td></tr>
<tr><td>34</td></tr>
<tr><td rowspan="3">E</td><td rowspan="3">P</td><td>16</td></tr>
<tr><td>19</td></tr>
<tr><td>22</td></tr>
<tr><td rowspan="2">F</td><td rowspan="2">N</td><td>23</td></tr>
<tr><td>21</td></tr>
<tr><td rowspan="2">G</td><td rowspan="2">M</td><td>20</td></tr>
<tr><td>14</td></tr>
<tr><td rowspan="6">Lecture</td><td rowspan="4">D</td><td>V</td><td>36</td></tr>
<tr><td>T</td><td>39</td></tr>
<tr><td rowspan="2">U</td><td>47</td></tr>
<tr><td>21</td></tr>
<tr><td rowspan="2">E</td><td rowspan="2">L</td><td>22</td></tr>
<tr><td>23</td></tr>
</table>

various process skills such as solving problems, working effectively with others, and finding trends in data.

Of the 44 items on the original SALG survey, 14 received a 70% or higher "n/a" response from more than one section. These items were removed from our analysis. The remaining 30 items had an overall response rate of greater than 95%.

For 28 of the 30 included items listed in Table V, POGIL students responded more positively about their course than lecture students. The notable exception to this trend is item 13 in Part A which indicates that POGIL sections found the text less helpful than lecture sections by almost one full point out of five. We attribute the low rating of the text to the fact that POGIL students use an activity book as their primary source of content in class every day and are therefore likely to also choose this book over the text as a general resource outside of class.

The SALG measures two distinct factors: Part A measures student perceptions of the value of course elements—15 items; and Part B measures perceived growth in process skills—15 items. The items in each part were summed for each student to generate a Part A score and a Part B score. These scores (not the item means reported in Table V) formed the basis for our analyses. By summing the Likert responses for the 15 questions in a given part of the survey we hoped to mitigate potential errors in the content validity of any one question. Missing values, which accounted for less than 5% of the responses, were replaced with the individual's mean response to that part.

Table V. SALG items and mean responses by method (N = sections per cohort)

SALG Part A Items	POGIL (N=9)		Lecture (N=6)	
"How much did the item help your learning"	*M*	*SD*	*M*	*SD*
1. The way in which the material was approached	3.88	0.45	3.44	0.80
2. How the activities/reading/assignments fit together	3.80	0.37	3.30	0.51
3. The pace at which we worked	3.60	0.35	3.02	0.45
4. Class presentations (including lectures)	3.63	0.52	3.56	0.86
5. Discussion in class	3.94	0.33	3.33	0.79
6. Use of in-class demonstrations or models	3.81	0.29	3.26	0.82
7. Opportunities for in-class review	3.50	0.48	3.15	0.91
8. The number and spacing of tests	3.65	0.45	3.26	0.61
9. The fairness of test content	3.48	0.45	3.24	0.62
10. The mental stretch required of us	3.77	0.42	3.33	0.49
11. The grading system used	3.63	0.57	3.19	0.65
12. The feedback we received	3.64	0.41	3.10	0.74
13. The text	2.44	0.51	3.35	0.84
14. The quality of contact with the teacher	3.98	0.40	3.65	0.91
15. The way this class was taught overall	3.93	0.50	3.54	0.78
Total score	54.68		49.7	

SALG Part B Items	POGIL (N=9)		Lecture (N=6)	
"Rate your growth in the listed item"	*M*	*SD*	*M*	*SD*
1. Working with peers outside of class	3.99	0.38	3.77	0.37
2. Solving problems	3.83	0.27	3.35	0.23
3. Finding trends in data	3.74	0.25	3.32	0.33
4. Working effectively with others	4.11	0.23	3.05	0.23
5. Communicating/explaining my ideas	3.86	0.33	3.06	0.27
6. Understanding the main concepts	3.91	0.38	3.65	0.38
7. Understanding relationship between concepts	3.76	0.42	3.56	0.41
8. Understanding how ideas in this class relate to other classes	3.49	0.35	3.37	0.42
9. Understanding the relevance of this field to real world issues	3.30	0.33	3.36	0.35
10. Appreciating this field	3.72	0.46	3.51	0.40
11. Ability to think through a problem or argument	3.83	0.35	3.43	0.32
12. Confidence in your ability to do this field	3.39	0.39	3.05	0.30
13. Feeling comfortable with complex ideas	3.57	0.34	3.20	0.24
14. Enthusiasm for subject	3.34	0.50	3.10	0.50
15. Remember/carry concepts to other classes	3.58	0.24	3.40	0.31
Total score	55.42		50.1	

Analysis of student scores by method showed that POGIL responses were significantly higher than lecture responses for both Part A and Part B. This analysis was done using a Mann-Whitney *U* test rather than an ANOVA since the data was not normally distributed. The results indicate that POGIL responses are significantly higher than lecture responses for both Part A ($z = -4.85$, $p<0.01$) and Part B ($z = -4.51$, $p <0.01$). Mean ranks (POGIL versus lecture) were 226.9 versus 170.7 for Part A, and 225.0 versus 172.8 for Part B. These values were used to calculate the effect sizes of teaching method, which were found to be moderate for both parts: Part A, $\eta^2 = .06$, and Part B, $\eta^2 = .05$ (*4*).

Discussion of SALG Survey Results

The results reported above indicate POGIL students perceived greater value in course offerings, and greater growth in their process skills, as compared to lecture students. However, this analysis ignores known significant differences between sections and instructors within a given cohort. As with the exam data, differences by section and instructor need to be explored and quantified in order to better assess what portion of the differences between the POGIL and lecture cohorts can be attributed to differences between the teaching methods.

To examine the possibility that student self-selection or random effects have caused a concentration of stronger students in the POGIL sections we analyzed (ANOVA) five independent measures of student aptitude among students at Institution D: Math SAT, Verbal SAT, General Chemistry 1 Grade, General Chemistry 2 Grade, and other course GPA in the semester of study. None of these variables showed statistically significant differences by method at the 95% confidence level. Of course, this does not address the possibility that there are other differences between POGIL and lecture sections due to self-selection based on learning styles preferences or other parameters not measured by the five variables listed above.

To evaluate the hypothesis that section, or the highly related variable of instructor, has a large effect on response scores within both the POGIL and lecture cohorts, analyses of variance in response scores by section and instructor were conducted within each cohort. The ANOVA results indicated significant effects on scores for both section and instructor in both cohorts. The mean effect size of section was large for Part A, $\eta^2 = .25$, and moderate for Part B, $\eta^2 = .12$. Similarly, mean effect size of instructor was large for Part A, $\eta^2 = .21$, and moderate for Part B, $\eta^2 = .07$ (*4*).

The moderate-to-large effect sizes of these variables opens the possibility that a few particularly strong sections or a few particularly effective instructors could, on their own, account for differences in response scores between POGIL and lecture. The former possibility (particularly strong POGIL sections) is not consistent with our analysis of sections at Institution D in which we found no evidence of variation in student aptitude between POGIL and lecture. The latter possibility (particularly effective POGIL instructors) led us to control our results using an independent measure of student perceptions of the course instructor: institutional end-of-course evaluation scores.

To test the hypothesis that POGIL sections appear more positive because the POGIL instructors are more effective than the lecture instructors, we evaluated response scores from sections taught by four instructors who received similarly high scores on institutional end-of-course evaluations (Instructors 1, 3, 5, 6). The results (ANOVA) showed no significant effect of method on scores for Part A, $F(1,216) = 2.81, p = .10$, but found method had a significant effect on scores for Part B $F(1,216) = 8.02, p < .01$. This indicates that, for the subset of students in sections that rated the instructor similarly by an independent measure, POGIL students did not report significantly higher contributions to their learning from the course elements found in Part A, but did perceive significantly greater gains in the process skills addressed by Part B.

Conclusions

In comparison to students enrolled in a lecture organic chemistry course, students enrolled in a POGIL organic chemistry course:

- Achieved higher scores on common exams including the comprehensive ACS Organic Exam. (An overview of exam results are reported in Table II. Mean ACS exam scores for Institution D [% correct out of 70] are reported by method and instructor in the bottom half of Table III.)
- Reported that elements of their course were at least as helpful to their learning (based on SALG Part A).
- Perceived greater gains in their own process skills (based on SALG Part B).

The latter two conclusions may in part explain the exam findings. That is, students who find course elements helpful and who build their process skills are, according to the learning theories underlying POGIL, more likely to succeed (*5,6*). The fact that POGIL students, as compared to lecture students, report that course elements are more valuable to their learning may be an outgrowth of the positive attitudes prevalent in POGIL courses. These positive attitudes are evident in the results of a survey of POGIL and lecture organic chemistry students which asked students to agree or disagree with the statement shown in Figure 1.

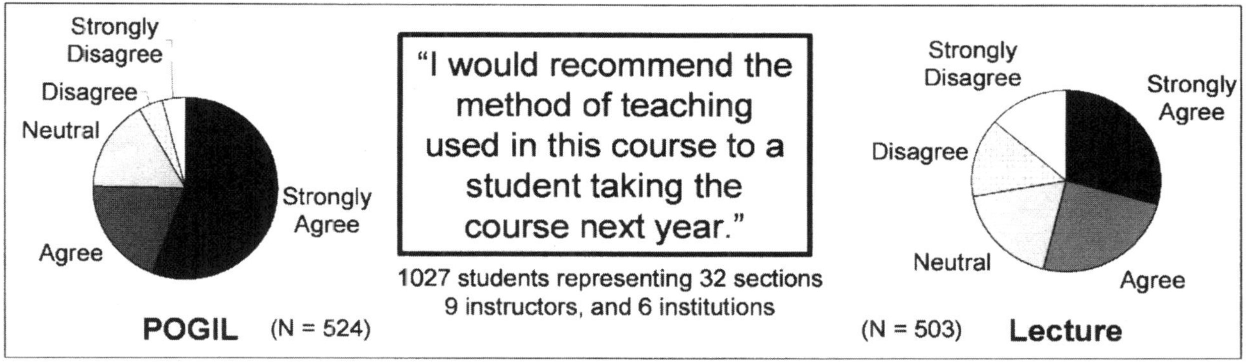

Figure 1. Survey of student attitudes toward POGIL and lecture

There were a total of 1027 responses (N_{POGIL} = 524, $N_{Lecture}$=503, N_{total} = 1027, response rate > 95%) from students at six institutions (B-G). Three-quarters of POGIL students agreed or strongly agreed with this statement, as compared to 54% of lecture students. At the negative end of the scale over a quarter of lecture students disagreed, while fewer than one in twelve POGIL students disagreed. Chi-square (χ^2) analysis of these data indicate that POGIL students are significantly more positive than lecture students regarding the method used in their classrooms (χ^2 = 102.48, p <.01).

The weight of all these conclusions is limited by uncertainty regarding the impact of the instructor, a key contributor to a student's experience in any course. Since there is evidence that high quality instruction was not limited to POGIL sections, it is unlikely that all the observed differences between POGIL and lecture survey scores are attributable to instructor. POGIL versus lecture studies that directly control for the variable of instructor are currently underway.

These studies provide evidence that POGIL may be more effective than lecture, but no explanation for why. An intriguing possibility is that some of the positive results reported here for POGIL may stem from the special opportunities POGIL affords instructors to improve their skill and knowledge as teachers. In a POGIL classroom instructors receive constant feedback while observing students discuss the material in their small groups. This information gives instructors unusually deep insight into student understandings and misunderstandings, as well as an opportunity to tailor teaching to individual learning styles. These dynamics could account for some of the observed attitudinal and exam score elevation among POGIL sections.

The above conclusions are consistent with studies of the effectiveness of POGIL in other contexts. For example, a study of POGIL involving 858 general chemistry students showed across the board improvement in exam scores, with a particularly significant drop in the DFW rate (*7*).

Student exam performance and student buy-in are both issues that deeply interest faculty considering change. We hope the data presented here help to inform the expectations of instructors interested in trying POGIL, and reassure their colleagues that a decision to use POGIL is a responsible one.

Anecdotally, we have found that while these data lead many faculty to try POGIL, their reasons for permanently switching to POGIL are often more personal. Some cite improvement in the quality of interactions with students; others find POGIL instruction less repetitive than lecture since it is student-centered. We find POGIL brings us closer to our ideal vision of a classroom: one that is filled with students actively engaged in discovery of a topic we love.

Acknowledgements

We would like to thank the students and professors involved in the study. Financial support was provided by the National Science Foundation POGIL Project (CCLI DUE 0231120), and the College of Charleston Summer Undergraduate Research with Faculty (SURF) Program.

References

1. Treisman, U. *College Mathematics Journal* **1992**, *23*, 362-372.
2. *The Carnegie Classification of Institutions of Higher Education,* URL http://www.carnegiefoundation.org/classifications/index.asp. Last accessed October, 2007.
3. Seymour, E. *Science Education* **2002**, *85*, 79-105.
4. Cohen, J. *Statistical power analysis for the behavioral sciences*; 2nd ed.; L. Erlbaum Associates: Hillsdale, N.J., 1988.
5. Lewis, S. E.; Lewis, J. E. *J. Chem. Educ.* **2005**, *82*, 135.
6. Zoller, U. *J. Res. Sci. Teach.* **1999**, *36*, 583-596.
7. Farrell, J. J.; Moog, R. S.; Spencer, J. N. *J. Chem. Educ.* **1999**, *76*, 570-574.

Chapter 20

Using an ACS General Chemistry Exam to Compare Traditional and POGIL Instruction

Martin D. Perry[1] and Randall D. Wight[2]

[1]Department of Chemistry, Patterson School of Natural Sciences, Ouachita Baptist University, 410 Ouachita Street, Arkadelphia, AR 71998
[2]Department of Psychology, School of Social Sciences, Ouachita Baptist University, 410 Ouachita Street, Arkadelphia, AR 71998

This study uses a standard ACS exam for general chemistry to examine the effectiveness of Process-Oriented Guided Inquiry Learning (POGIL) instruction compared to traditional instruction. One cohort in this general chemistry study received a full year of POGIL instruction, while the second cohort received one semester of traditional instruction followed by a semester of POGIL instruction. The ACS General Chemistry Full-Year Exam was administered to both cohorts at the end of the second semester and statistical comparisons were made. The data indicated no statistical differences in overall exam performance; however, closer examination of individual questions did yield insights into the learning process.

The Process-Oriented Guided Inquiry Learning (POGIL) instructional method, described in detail elsewhere in this text and other sources (*1-4*), aims to improve students' process skills in addition to assisting them with mastering course content. The key process skills are information processing, critical thinking, problem solving, communication, teamwork, management, and assessment. Instructors often suggest a variety of reasons for not attempting this student-centered approach where learners work in self-managed teams. Chief among them is the notion that course content or coverage of material will be lacking (*5*), leading thus to poor performance on standardized exams such as those published by the American Chemical Society (ACS) Examinations Institute or to inadequate preparation for future courses including graduate and professional training.

Achievement in chemistry courses is often linked to performance on ACS standardized exams (*6*). Although these exams provide a benchmark for comparison and set the standards deemed appropriate by the profession, they are but one tool for assessing student learning. This study explores instructional method as a tool and uses a standard ACS exam for general chemistry to compare the effectiveness of POGIL and traditional instruction.

Methodology

All students were enrolled in General Chemistry courses at a small liberal arts university located in the rural mid-South. During the 2001-02 academic year (2002 cohort), every student experienced POGIL instruction both semesters by the same instructor. During the 2002-03 academic year (2003 cohort), instruction ranged from traditional during the first semester and POGIL during the second. A different instructor taught each semester. Prior academic preparation of the two cohorts—measured by ACT scores (2002, $N = 26$, $M = 27.4$, $SD = 3.31$; 2003, $N = 34$, $M = 27.2$, $SD = 3.12$) and high school grade point averages (2002, $N = 26$, $M = 3.73$, $SD = .360$; 2003, $N = 34$, $M = 3.79$, $SD = .208$)—exhibited no statistical differences, respectively, $F(1, 58) = .08$, $p = .77$, $d = .06$ and $F(1, 58) = .65$, $p = .43$, $d = .24$. Thus, the only discernable preparatory difference between the cohorts was the instruction used in the first semester. The 1999 General Chemistry Full-Year Exam (*7*) was administered to both cohorts at the end of the second semester as the final exam for the course sequence.

Materials used for POGIL instruction were authored by Moog and Farrell (*8*). Each class period began with a brief (five min) question-and-answer period, followed by students working in self-managed teams of three to four students on the assigned activity. Exercises requiring students to apply concepts invented during class time were assigned and collected at the beginning of the next class meeting. The students were allowed to work with each other or seek assistance from the instructor between class sessions. Occasionally, quizzes were given in lieu of turning in the assigned exercises. In addition to the final exam, four exams were administered during the course of the semester. The exam format consisted of multiple choice, short answer, brief discussion, and algorithmic problem solving questions. While the class sessions were conducted using POGIL methods, laboratories for all cohorts were under the purview of a separate instructor who used a traditionally structured manual.

Results

We began our analysis by comparing the average percent correct for local (M = 64.28, SD = 21.56) and national (M = 59.89, SD = 16.63) students for the two cohorts (N = 140 question pairs). A Pearson product-moment revealed a substantial positive relationship between the two percentage samples, $r(209)$ = .745, p < .001 (two-tailed). Though the local sample contained greater variability, the local mean was significantly higher than the national mean, $t(210)$ = 4.42, p < .001, d = .23. A comparison of the overall percent correct on the ACS questions (N = 70) revealed no statistical difference—$F(1,138)$ = 1.58, p = .21, d = .21—for the two local cohorts: 2002 (M = 65.88, SD = 21.19) and 2003 (M = 61.26, SD = 22.26). Figure 1 displays the difference for each item (2002 score minus 2003 score) in local cohort scores: positive differences indicate questions where the 2002 cohort scored higher with negative difference indicating the opposite. The 2002 cohort achieved greater percent correct values on 40 of the 70 items overall.

To assess the efficacy of traditional instruction compared to POGIL instruction, we focused on Semester 1 where instruction type varied for the two local cohorts. Questions from the ACS General Chemistry Full-Year Exam were examined and assigned to Semester 1 (N = 35) or Semester 2 (N = 30) according to presentation of material. Five of the questions received no coverage in either semester. When comparing percent correct on the germane national questions, the 2002 (POGIL) cohort (M = 67.36, SD = 19.22, N = 35) and the 2003 (traditional) cohort (M = 60.59, SD = 23.52, N = 35) did not differ significantly, $F(1, 68)$ = 1.73, p = .19, d = .31. The 2002 cohort achieved greater percent correct values on 21 of the 35 Semester 1 items.

Although statistically significant differences in ACS exam performance do not exist between the two cohorts overall or by individual semester, eight questions (see Table I) do differ significantly, with the 2002 cohort performing better on six of the eight. Additionally, six of the eight questions were first-semester topics where the cohorts received different methods of instruction. The 2002 cohort scored higher on topics such as molecular structure (bond order and bond angles), stoichiometry (limiting *and* excess reagents), equilibrium (graphical interpretation), and experimental (graphical interpretation and safety).

Of the five exam questions (questions 42, 60, 62, 63, & 64) that did not receive any coverage during either semester for either cohort, the 2002 cohort scored higher on four of them (questions 42, 60, 62, & 64). Additionally, fifteen questions on the exam (questions 8, 14, 15, 20, 31, 44, 47, 49, 51, 55, 60, 61, 62, 66, & 68) yielded mean correct values (see Table I) that were 1.50 or higher for both cohorts signifying that both cohorts failed to answer above 50% correct. In most cases, these questions required generalizations or higher-order levels of analysis in order to determine the correct answer. Ten of these tougher questions

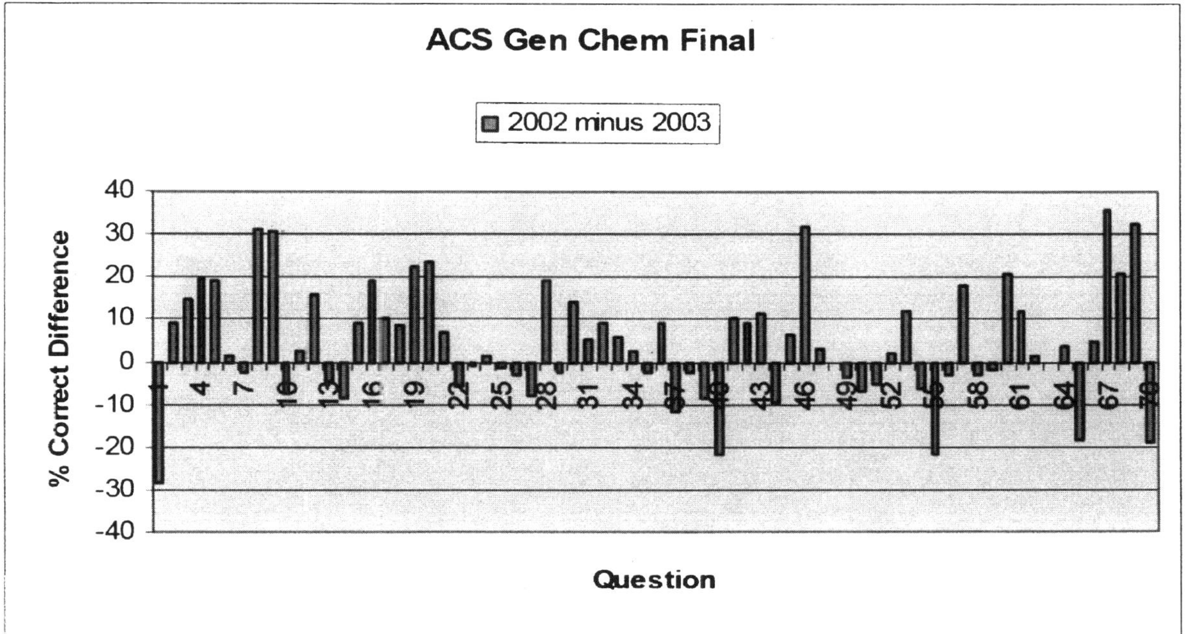

Figure 1. Percent correct difference for each of the 70 items on the ACS exam. Positive differences indicate questions where the 2002 cohort performed better. Negative differences indicate questions where the 2003 cohort performed better.

Table I. Mean Correct[a], Standard Deviations, and One-Way Analysis of Variance (ANOVA) Comparing Cohorts on ACS Standardized Questions

Question (Semester Coverage)	2002 ($N = 26$)		2003 ($N = 34$)		ANOVA
	M	*SD*	*M*	*SD*	$F(2,81)$
1 (1)	1.58	.504	1.29	.462	4.48**
2 (1)	1.12	.326	1.21	.410	1.72
3 (1)	1.38	.496	1.53	.507	1.25
4 (1)	1.42	.504	1.62	.493	2.00
5 (1)	1.19	.402	1.38	.493	2.91
6 (1)	1.08	.272	1.09	.288	.18
7 (1)	1.35	.485	1.32	.475	.29
8 (1)	1.54	.508	1.85	.359	4.87**
9 (1)	1.19	.402	1.50	.508	3.37***
10 (2)	1.19	.402	1.12	.327	.85
11 (1)	1.27	.452	1.29	.462	.27
12 (1)	1.46	.508	1.62	.493	.92
13 (1)	1.08	.272	1.03	.171	.37
14 (1)	1.62	.496	1.53	.507	2.10
15 (1)	1.62	.496	1.71	.462	1.83
16 (1)	1.19	.402	1.38	.493	2.16
17 (1)	1.08	.272	1.18	.387	2.78
18 (1)	1.27	.452	1.35	.485	.24
19 (1)	1.42	.504	1.65	.485	2.12
20 (1)	1.50	.510	1.74	.448	2.85
21 (1)	1.08	.272	1.15	.359	3.10***
22 (1)	1.12	.326	1.06	.239	1.50
23 (2)	1.04	.196	1.03	.171	.47
24 (2)	1.31	.471	1.32	.475	.38
25 (1)	1.42	.504	1.41	.500	.06
26 (1)	1.38	.496	1.35	.485	2.49
27 (1)	1.46	.508	1.38	.493	.43
28 (1)	1.19	.402	1.38	.493	1.28
29 (1)	1.12	.326	1.09	.288	.09
30 (2)	1.27	.452	1.41	.500	.66
31 (2)	1.54	.508	1.59	.500	.47
32 (2)	1.23	.430	1.32	.475	.95
33 (2)	1.00	.000	1.06	.239	1.02
34 (2)	1.38	.496	1.41	.500	.18
35 (2)	1.23	.430	1.21	.410	1.06
36 (2)	1.12	.326	1.21	.410	.76
37 (2)	1.50	.510	1.38	.493	1.66
38 (1)	1.12	.326	1.09	.288	.19
39 (1)	1.50	.510	1.41	.500	.26

Table I. *Continued.*

Question (Semester Coverage)	2002 ($N = 26$)		2003 ($N = 34$)		ANOVA
	M	*SD*	*M*	*SD*	$F(2,81)$
40 (2)	1.31	.471	1.09	.288	2.79
41 (1)	1.08	.272	1.18	.387	.91
42 (0)	1.23	.430	1.32	.475	.36
43 (2)	1.15	.368	1.26	.448	.75
44 (2)	1.65	.485	1.56	.504	1.43
45 (2)	1.35	.485	1.41	.500	.80
46 (2)	1.15	.368	1.47	.507	3.63***
47 (2)	1.50	.510	1.53	.507	1.79
48 (2)	1.27	.452	1.26	.448	2.77
49 (2)	1.65	.485	1.62	.493	2.50
50 (2)	1.42	.504	1.35	.485	.24
51 (2)	1.73	.452	1.68	.475	1.04
52 (2)	1.04	.196	1.06	.239	.70
53 (2)	1.12	.326	1.24	.431	.73
54 (2)	1.38	.496	1.32	.475	.25
55 (2)	1.81	.402	1.59	.500	1.97
56 (2)	1.50	.510	1.47	.507	.42
57 (2)	1.12	.326	1.29	.462	3.01
58 (1)	1.38	.496	1.35	.485	3.28***
59 (1)	1.08	.272	1.06	.239	.88
60 (0)	1.50	.510	1.71	.462	1.78
61 (1)	1.62	.496	1.74	.448	1.70
62 (0)	1.69	.471	1.71	.462	.94
63 (0)	1.27	.452	1.26	.448	.01
64 (0)	1.35	.485	1.38	.493	.08
65 (1)	1.54	.508	1.35	.485	1.03
66 (2)	1.92	.272	1.97	.171	.97
67 (2)	1.12	.326	1.47	.507	4.85***
68 (1)	1.62	.496	1.82	.387	2.27
69 (1)	1.38	.496	1.71	.462	4.16**
70 (2)	1.42	.504	1.24	.431	1.29

[a] where 1 = Correct, 2 = Incorrect

** $p < .01$

*** $p < .05$

(questions 8, 15, 20, 31, 47, 60, 61, 62, 66, & 68) were handled more aptly by the 2002 cohort with an average percent correct 13.1% higher than the 2003 cohort.

Discussion

The statistical analysis reveals that both local cohorts scored significantly above the national average and that no differences existed between local cohorts. These data contradict suggestions that less content coverage or less instructor control would result in lower standardized exam scores. Even though the 2002 cohort covered approximately 85% of course material compared to the 2003 cohort, their overall performance was not significantly different. While small sample sizes and large standard deviations contributed to this finding, the 2002 cohort did average 4.5% correct higher exam scores.

Because the only difference in instruction for the two cohorts occurred during the first semester of the course, a second analysis of the 35 first-semester questions on the ACS exam was conducted. Similarly, the results exhibited no statistical differences. However, when individual questions are examined, eight questions show significant differences including six first-semester questions where the 2002 cohort scored higher. Further examination of these questions revealed that these topics required higher-order reasoning, multiple steps, or data interpretation. These data suggest that key process skills are developed to a greater extent in the POGIL classroom.

The final analyses continue to bear out this trend. The 2002 cohort's greater performance on no-coverage and tougher questions highlights their reliance on skills developed during class sessions. As the students work in self-managed teams, they are forced to grapple with new ideas, invent concepts, assess their learning, and communicate with their peers. Rarely do any of these practices occur within the framework of traditional classrooms. Without the practice needed to acquire these grappling skills, students may flounder when confronted with tough questions or novel ideas. Instructors bear the burden of equipping students with the tools needed for success, not just on exams, but in careers and life.

References

1. Farrell, J. J.; Moog, R. S.; Spencer, J. N. *J. Chem. Educ.* **1999**, *76*, 570-574.
2. Hanson, D.; Wolfskill, T. *J. Chem. Educ.* **2000**, *77*, 120-130.
3. Hinde, R. J.; Kovac, J. *J. Chem. Educ.* **2001**, *78*, 93-99.
4. Lewis, J. E.; Lewis, S. E. *J. Chem. Educ.* **2005**, *82*, 135-139.

5. Williams, G.; McTighe, J. *Understanding by Design*; ASCD: Alexandria, VA, 1998.
6. Holme, T. *J. Chem. Educ.* **2003**, *80*, 594.
7. *General Chemistry Full-Year Exam*, ACS Examinations Institute: Clemson, SC, 1999.
8. Moog, R.S.; Farrell, J.J. *Chemistry: A Guided Inquiry, 3rd Edition*; John Wiley & Sons: New York, 2006.

Indexes

Author Index

Subject Index

B

C

H

J

K

L

P